IMAGES
of America

HAMDEN FIREFIGHTING

Hamden Fire Chiefs

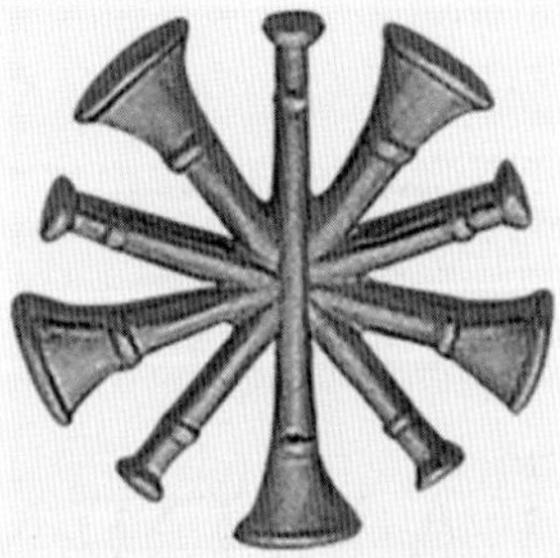

Charles Loller
1925 - 1942

Raymond C. Spencer
1942 - 1960

V. Paul Leddy
1960 - 1984

John Tramontano
1984 - 1993

Paul Wetmore Sr.
1993 - 1996

Timothy Sullivan
1996 - 2002

James Leddy
2002 - 2006

David Berardesca
2006 -

When the Hamden Board of Fire Commissioners was created by the state legislature in 1925, its first act was to appoint Charles Loller as Hamden's first fire chief. Chief Loller's successors, Chiefs Raymond C. Spencer and V. Paul Leddy, were also commission appointees. After a 1966 charter revision, Hamden's mayor became the appointing authority for all future chiefs, whose appointments were confirmed by the Hamden Legislative Council. (Hamden Fire Retirees Association.)

On the Cover: Parked on the ramp of Whitneyville Station 3 in May 1939 are Hamden's "Emergency Squad" (left), on a 1938 Diamond-T chassis, and Engine 2, a 1938 Seagrave Canopy Cab 600-gallons-per-minute (gpm) pumper. Personnel are, from left to right, Firefighter Roland Ruwet, Assistant Chief Raymond C. Spencer, and Firefighters Joseph Dukat and Robert Reutenauer. (Hamden Fire Retirees Association.)

IMAGES
of America

HAMDEN FIREFIGHTING

David Gray Johnson
Foreword by Chief David A. Berardesca

ARCADIA
PUBLISHING

ISBN 978-1-4671-2761-5

Published by Arcadia Publishing
Charleston, South Carolina

Printed in the United States of America

Library of Congress Control Number: 2017941856

For all general information, please contact Arcadia Publishing:
Telephone 843-853-2070
Fax 843-853-0044
E-mail sales@arcadiapublishing.com
For customer service and orders:
Toll-Free 1-888-313-2665

Visit us on the Internet at www.arcadiapublishing.com

For Barbara, and those dedicated, courageous, funny, inspiring, and caring people I was honored to work with on the department—you will always be remembered

Contents

Foreword 6

Acknowledgments 7

Introduction 8

1. The Beginning 11

2. The Hamden Fire Department 27

3. The "Paid" Department 41

4. Expansion 65

5. Moving Forward 103

6. A New Century 119

Foreword

There are many experiences in a firefighter's career that can be considered defining moments. My career-defining moment was most certainly my appointment as chief of the Hamden Fire Department. This is an honor that has also been bestowed upon seven before me. Each of these gentlemen was tasked with the responsibility of taking his department to the next level while honoring the traditions that make up the unique history of our treasured Hamden Fire Department.

As you read through the following pages, you will come to understand the challenges faced by those before us and how those challenges helped to create who we are today. The tasks and challenges for me and for those who follow are spelled out in our history—that is, work to make a better and safer future for citizens and firefighters by the lessons learned from our history.

Through the efforts of Capt. David Gray Johnson (retired), our unique history has been painstakingly preserved in the pages that follow. Captain Johnson will guide you through the origins and challenges of our volunteer firefighting forces to the modern professional department we enjoy today.

In closing, many a firefighter has stated, "I could write a book about this place." Well, that book has been written, and it is with pride that I am able to be part of this historically rich organization. Thank you to all members of the Hamden Fire Department, past, present and future.

—Chief David A. Berardesca
Hamden Fire Department

Acknowledgments

A pictorial history on any subject is a collaborative effort by many people—not just the photographers who shoot the photographs and the scribes who provide the accounts but also those individuals who have the presence of mind to preserve those images and accounts for posterity. *Hamden Firefighting* would not have been possible without those keepers of the flame, so to speak, to whom I am deeply grateful.

Many of the images included herein, captured by noted area photographers Lewis Berlepsch, George Keeley, Michael Kurtz, John Mongillo Sr., John Mongillo Jr., Edward Quimby, I.A. Sneiderman, Calvin Stoner, Arthur J. Tefft, Domenic Vaccaro, and Donald Yosua, came from a collection compiled over several decades by what is now the Hamden Fire Retirees Association. Others were generously shared by Chief Paul Wetmore (retired), Deputy Chief Gary Merwede, Battalion Chiefs Thomas Doherty and Gilbert Spencer (both retired), Firefighter Richard Carney (retired), Firefighter Brian Leddy, Chandler Brainard, Lawrence Gershman, and Sally Hines-Row, with the help of *New Haven Register* publisher Kevin Corrado and the Hamden Historical Society.

Additional photographs were obtained from the John DellaVecchia Collection in the Hamden Historical Society, Matthew Lee, Joseph Pepe, Beth Shutts, Chief Timothy Sullivan (retired), Joseph Taylor, Erwin Wetmore, and the late G. Donald Steele.

Also included are outstanding photographs by Chief John Tramontano (deceased), Lt. Robert Mordecai (retired), Lt. Jeffrey Pechmann, Firefighters John Corbett and Raymond Dobbs (both retired), Joseph Ciscone, Kenneth DeLorme, Edward Doiron Jr., Jack Harriett, George Moehl, Leland Robinson, and Harold Wheatley; all of them provided important contributions to this book.

I want to thank Russell Loller, who, over 40 year ago, permitted me to copy the personal notes of his great-grandfather Chief Charles Loller.

Finally, I am especially grateful to my fellow firefighter and close friend of more than 50 years, Firefighter Edward Doiron Sr. (retired), for the hundreds of excellent Hamden Fire Department photographs he has taken through the years. No one has done more to make *Hamden Firefighting* possible.

When known, the name of the photographer or original publication appears in parentheses with each photograph caption. Otherwise, photographs are attributed through the courtesy of the organizations or individuals who provided them. Hamden Historical Society is abbreviated as HHS, and Hamden Fire Retirees Association is HFRA.

—Capt. David Gray Johnson
Hamden Fire Department (Ret.)

INTRODUCTION

Today, the Hamden Fire Department is a proud, highly professional, state-of-the-art guardian service of 100 career and roughly two dozen volunteer personnel, protecting over 60,000 residents. In 2016, the department answered over 10,000 calls, 74 percent of which were for emergency medical services. It took many decades for Hamden's fire service to arrive here, and it got a late start.

On July 1, 1853, when a fire heavily damaged the roof and destroyed the contents of Mount Carmel's "fire proof" Hamden Iron Works, the *New Haven Palladium* reported that the loss was covered by insurance, but there was no mention of any effort to extinguish the blaze. In September 1875, an early morning fire in Centerville started in the J.T. Henry Company and extended to the New Haven Web Company, destroying several buildings. The *New Haven Register* noted, "The fire was under good headway when discovered, as there was no engine or other apparatus with which to combat the flames, the building with its contents was soon destroyed."

Within months, the charred ruins of the New Haven Web Company were replaced by a voluminous three-story brick structure later known as the Web Shop. The Web Shop, which was enlarged several times in the 65 years that it stood on Whitney Avenue at Mill River, was Hamden's largest employer. The huge complex's only fire protection was numerous fire extinguishers throughout and a sprinkler system fed by an enormous water tower at the rear of the main building. That was the extent of Hamden's fire defenses until the late 19th century.

It was not until the mid-1890s, after a spate of suspicious fires destroyed barns in southern Hamden, that a group of Highwood residents finally decided that it was time to organize Hamden's first fire company, Highwood's Pioneer Hose Co. No. 1, on December 10, 1896.

During the next 15 years, residents in Centerville, Hamden Plains, Whitneyville, Mount Carmel, and North Hamden organized fire companies in their districts. Fire companies in the State Street, Mix District, and Dunbar Hill areas all followed during the 1920s. As late as 1956, residents in West Woods organized their own volunteer company, which served their district for 32 years.

A now long-forgotten volunteer fire company was the reason why the Mt. Carmel Volunteer Fire Co., organized on November 8, 1911, added a numerical designation to its official name. The company's officially chartered name is Mt. Carmel Volunteer Fire Company, No. 1 Incorporated. The "No. 1" was included because three weeks after the fire company was organized in St. Mary's Hall, the North Hamden Improvement Society organized a volunteer fire company in the Liberty Cartridge Company building, formerly occupied by the Mount Carmel (Upper) Axle Works. A.J. Ralph was elected the company's first chairman, and Judge Willis Cook was its first chief.

Located on Whitney Avenue opposite West Woods Road, Liberty Cartridge was in an area once called "Mount Carmel Center," near Mount Carmel Avenue and the Kimberly Store. The new volunteer fire company purchased Liberty Cartridge's hose cart, 400 feet of fire hose, and playpipe for $220.60. After the New Haven Water Company acquired the Liberty Cartridge building in 1914, the North Hamden Improvement Society met in various local homes and at the Mount Carmel Congregational Church parish house.

With the passage of time, the North Hamden fire company became nothing more than an unconfirmed legend because no records of meetings, fires, or photographs of company members, meetings, or any other activities could be found. Then, several years ago, Battalion Chief Gilbert Spencer (retired) found minutes of two 1914 meetings in the Hamden Historical Society. In 2014, all meeting minutes and many other company records of the North Hamden Improvement Society were discovered by local historian and writer Mary Jane McGaffin (now deceased), who turned them over to the Hamden Fire Retirees Association.

The North Hamden Improvement Society disbanded in 1926, its assets and membership being absorbed into the Mt. Carmel Volunteer Fire Co.

It was not until 1925 that the Hamden Board of Fire Commissioners was created by an act of the state legislature. The various independent volunteer fire companies were each assigned a number and were now under the control of a central municipal authority, the Hamden Fire Department, headed by Chief Charles Loller.

In an era when few people operated motor vehicles of any kind, the advent of motorized fire apparatus created the need for firefighters who were not only skilled drivers but also capable of operating the chemical tanks and pumps. The so-called paid department began with the paid drivers who, as early as 1918, were elected from the ranks of the individual volunteer fire companies that initially paid their salaries. It was the driver's job to get the apparatus to the fires, and with his help, it was the job of the volunteer firefighters to do the rest.

The Hamden paid driver of the 1920s was married to his job. He worked 24 hours a day, seven days a week, with time off only when he paid a substitute to take his place. Hamden's first paid driver, Charles Coe, a founding member of the Humphrey Hook and Ladder Association, was paid $359 by that company in 1918 to be the janitor. A year later, he was driver of its new 1919 Seagrave pumper.

As Hamden's population grew and more industries came to town, the number of paid personnel expanded. The department was reorganized in 1942 under a paid chief, with two paid company officers on each of two platoons of 11 men, working an 84-hour week. The postwar years saw the department shift from a largely volunteer to a predominantly paid, or "career," force that had more than doubled in size. Between 1948 and the late 1950, the firefighters' workweek dropped from 84 hours to 67.1 hours to 56 hours, when a third platoon was added.

The greatest expansion in the department's size and scope of responsibility occurred during the nearly 24-year tenure of Chief V. Paul Leddy (1960–1984). By 1970, two new fire stations were built, one new engine and one new truck company were added, and line personnel, now on four platoons, were on a 42-hour workweek. Peaking at 124 officers and firefighters in 1980, Hamden was the third-largest career fire department in New Haven County, smaller than only the Waterbury and New Haven departments. Only four volunteer companies remained, but each still had a thriving and active membership. It was during Chief Leddy's 24 years that emergency medical services became a major part of Hamden's fire service, which today accounts for nearly three quarters of its responses.

Hamden Firefighting contains many historical images from gifted local photographers, including Arthur Tefft, I.A. Sneiderman, John Mongillo Sr., John Mongillo Jr., and many others, whose photographs help to tell the story. These wonderful images have been preserved by the Hamden Historical Society, the Hamden Fire Retirees Association (HFRA), and in several private collections. Chief John Tramontano (now deceased), Battalion Chief Gilbert Spencer (retired), Battalion Chief Thomas Doherty (retired), and Chandler Brainard have preserved and donated literally hundreds of department photographs and newspaper articles to the HFRA archives.

Selecting and compiling the photographs for *Hamden Firefighting* was not an easy task. Space was a limiting factor, of course, but there was also the matter of the missing photographs. Photographs of almost every early piece of apparatus and fire company and most of the major fires are included here. Unfortunately, there are several major emergencies, some of them relatively recent, for which no original photographic prints were available.

The earliest Hamden fire for which photographs are known to exist is the 1918 blaze that destroyed the old Hamden Plains Methodist Church. A photograph of the aftermath is in this book. But original prints of the actual fire could not be found—only inferior small-scale halftone reprints that were not suitable. Other major fires of the early 20th century, such as the 1921 Economy Concrete fire that resulted in a $100,000 loss or the $165,000 Dextone fire at the same location in 1932, may not have been photographed at all.

More recently, a gas explosion leveled a Whitneyville apartment building in October 1977. Several elderly residents were rescued by Firefighter Ralph Tomaselli of Hamden (at the time off duty) and some friends, who were nearby when it happened. We could find no glossies of that incident, or any original photographs of the 1978 Daw Packing Company fire, the Corey Lumber Company and the Howard Johnson's Restaurant fires on the same night in April 1979, or the Rondo fire later that year.

The June 1982 flood, when so many Hamden career and volunteer firefighters worked around the clock to rescue countless residents, some from very perilous situations, is documented here only with Leland Robinson's stunning photograph of floodwaters encroaching on the covered bridge near the Lake Whitney Dam.

But we are extremely grateful for those original photographs and accounts that have been preserved so they could be presented here, because historic preservation is what *Hamden Firefighting* really is all about.

In December 1984, the *Hamden Chronicle* ran a story commemorating the 20th anniversary of two legendary Hamden restaurant fires that occurred within weeks of each other. The article concluded with a request for readers to contact fire headquarters should they have any fire department photographs or memorabilia they might want to share. One reader stopped by the chief's office a few days later and donated a book she found at the curb years earlier during one of Hamden's seasonal trash pickups. It was the official Hamden Fire Department log book, detailing every emergency call from 1931 to 1941.

The woman's discovery was both fortuitous and troubling. The fortuitous part is obvious. The troubling part was the fact that the book was discarded in the first place.

Preserving history, especially that of a municipal guardian service, is a sacred mission. It is hoped that *Hamden Firefighting* will serve not only to inform the reader about some of the rich history of the Hamden Fire Department but also to inspire others everywhere to keep their histories alive by preserving the photographs and narratives that tell their stories. Perhaps, like the lady who dropped off that old Hamden Fire Department logbook at the chief's office in 1984, some local readers of *Hamden Firefighting* will discover photographic gems from our past in their collections that we may gratefully scan for *Hamden Firefighting II*.

Hamden Firefighting is a labor of love that celebrates a special group of individuals, past and present, career and volunteer, whose mission for over 120 years has been to protect the people who live and work in our town.

Note: Certain nomenclature changes unique to this department require some explaining. At various times since 1983, the title for the department's second-in-command has been either *assistant chief* or *deputy chief*; it is presently the latter. Since 1942, department shift commanders have had the titles *captain* (1942–1954), *battalion chief* (1954–1961), *deputy chief* (1961–1984), *commander* (1984–1995), and again, *battalion chief* (1995–present). In the captions that accompany the photographs, the titles of the second-in-command chiefs and the shift commanders are whatever they were on the date of the photograph. Retired shift commanders who once had the title of deputy chief will always be identified as such. Retirees promoted to shift commander after 1983 will always be identified as retired battalion chief.

One

The Beginning

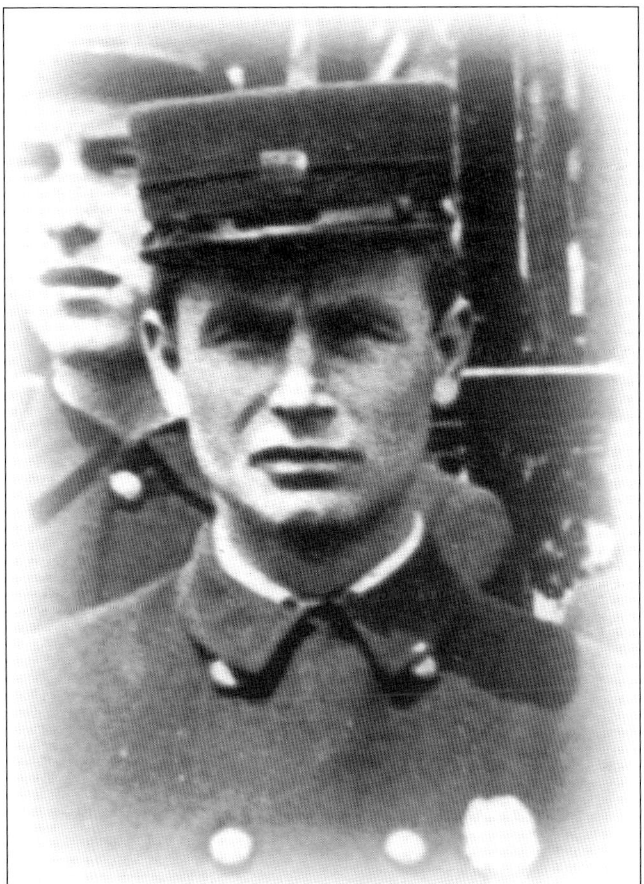

Charles Coe was a founding member of the Humphrey Hook and Ladder Association. In 1918, he was hired by the fire company to be its "janitor." When the town purchased a new Seagrave pumper the following year for the Humphrey station, Coe became the department's first paid driver. (Humphrey Hook and Ladder Association.)

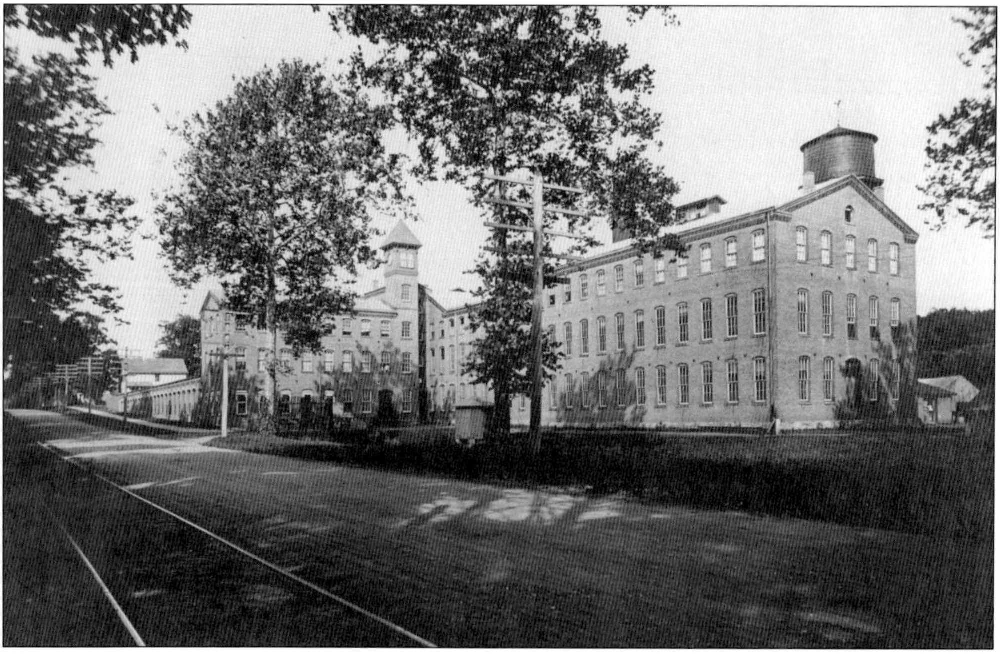

The Web Shop was once Hamden's largest employer. Located on the east side of Whitney Avenue, exactly where the Wilbur Cross Parkway crosses today, this rambling complex was built after the several wood-frame buildings on that site burned to the ground in September 1875 because there was no organized fire protection in Centerville or anywhere else in Hamden. For the 65 years that this building stood before being razed for the parkway in 1940, initial fire protection was provided by an internal sprinkler system. The water tank in this 1904 photograph supplied the sprinkler system until the building was connected to the public water system in the 1910s. Below, in one of the few existing interior views of the Web Shop, sprinkler heads in this 1912 photograph can be seen in the feeder pipes that run along the ceiling. (Both, Joseph Taylor.)

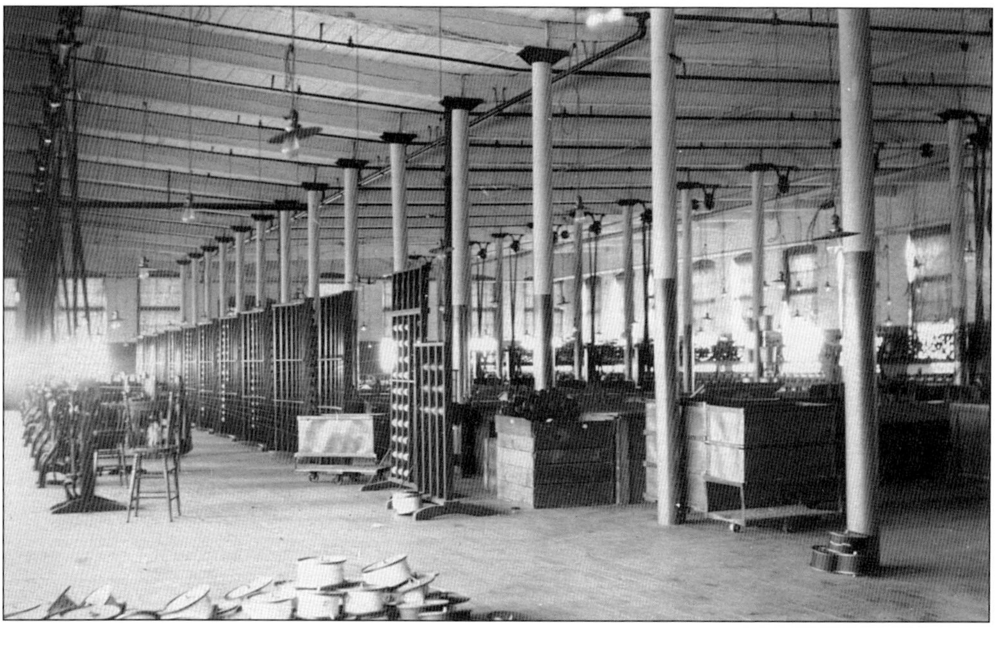

$100 REWARD.

WANTED FOR BARN BURNING IN HAMDEN, CONN.

WESLEY PORTER, white; about 28 years old; about 5 feet 6 inches hight; weight about 145 lbs.; light sandy hair rather thin; small sandy moustache; small irregular tufts of sandy hair in various parts of face; (may now be shaved clean) blue eyes; rather long face and very red nose; of very quiet disposition, talks very little except when under the influence of liquor.

When last seen, July 30, about 3 p.m. he wore a black Derby but well worn; a dirty brown woolen or partly woolen shirt; an old pair of suspenders; a pair of half worn out dark colored pants; a pair of well worn gaiters. Had been engaged as a farm hand by Mr. Samuel A. Flight of Hamden, Conn. Claimed to have been employed just previously by Sperry & Barnes of New Haven, Conn., and to hail from Massachusetts.

In the afternoon of Wednesday, July 30, 1890, Wesley Porter while under the influence of liquor had an altercation with his employer Mr. Sam'l A. Flight, who discharged him. Out of revenge Porter then set fire to his employer's barn and immediately fled, leaving all his things, except what he happened to wear, behind him.

Mr. Flight will pay a reward of $100. for the capture of said Porter. Send all information concerning said Porter to said Samuel A. Flight, P. O. Box 74, Highwood, Conn., or to,

CHARLES F. BOLLMANN,
Sup't of Police
New Haven, Conn.

July 31, 1890

Numerous suspicious barn fires in the mid-1890s inspired a group of Highwood residents to meet in William Dicks's barbershop to discuss the formation of Hamden's first fire company. On December 10, 1896, the citizens met again in Visel's Hall on Morse Street to organize the Highwood Volunteer Fire Association. (John DellaVecchia Collection, HHS.)

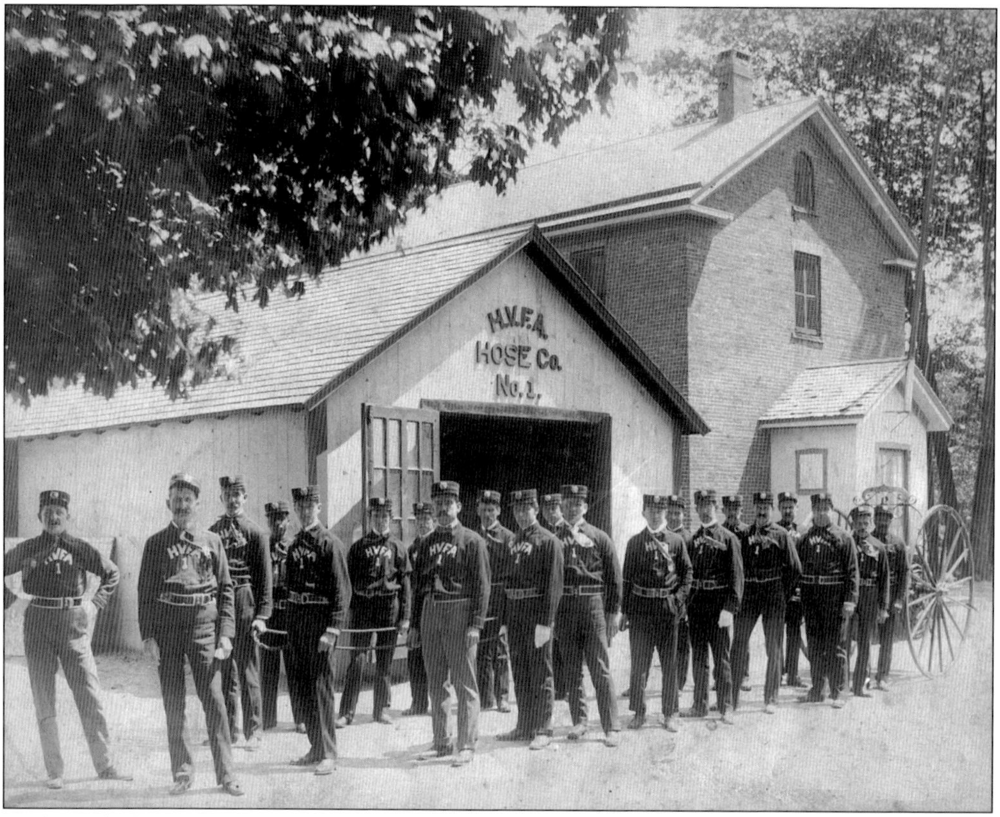

Highwood's fire company members at their first quarters on Alstrum Street are, from left to right, Charles Pellert, Samuel Williams, ? Doepensmith, George Madison (or maybe Joseph Jackson), John Mautte, three unidentified men, Thomas O'Connell, George Visel, Paul Ert, David F. Howe, Robert O'Connell, two unidentified men, George Moore, Charles Loller, James Rogers, Dennis Keeley, Frank Butler, and Charles Groves. (John DellaVecchia Collection, HHS.)

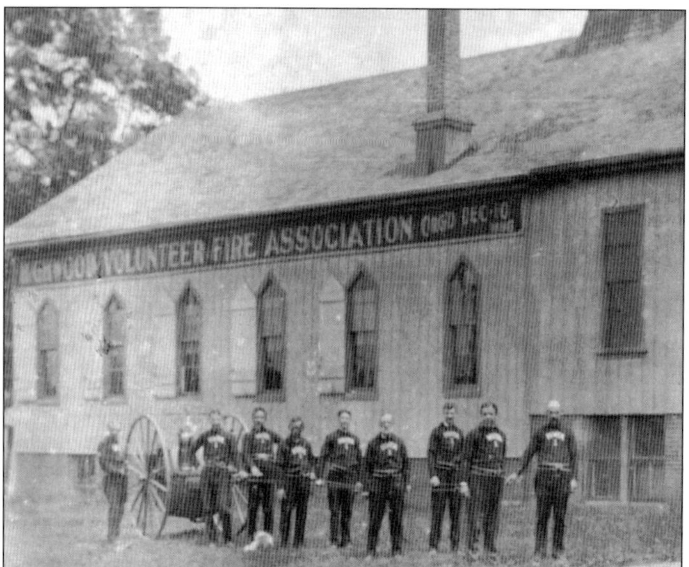

In 1898, the Highwood volunteers moved their quarters to the New Lebanon Mission on Morse Street. In 1909, they built their firehouse at Dixwell Avenue and Morse Street. The building was sold when Highwood's apparatus moved to Humphrey Station 2 in 1951. A home to many businesses since, the building remains standing in 2017. The New Lebanon Mission building was razed in 1974. (John DellaVecchia Collection, HHS.)

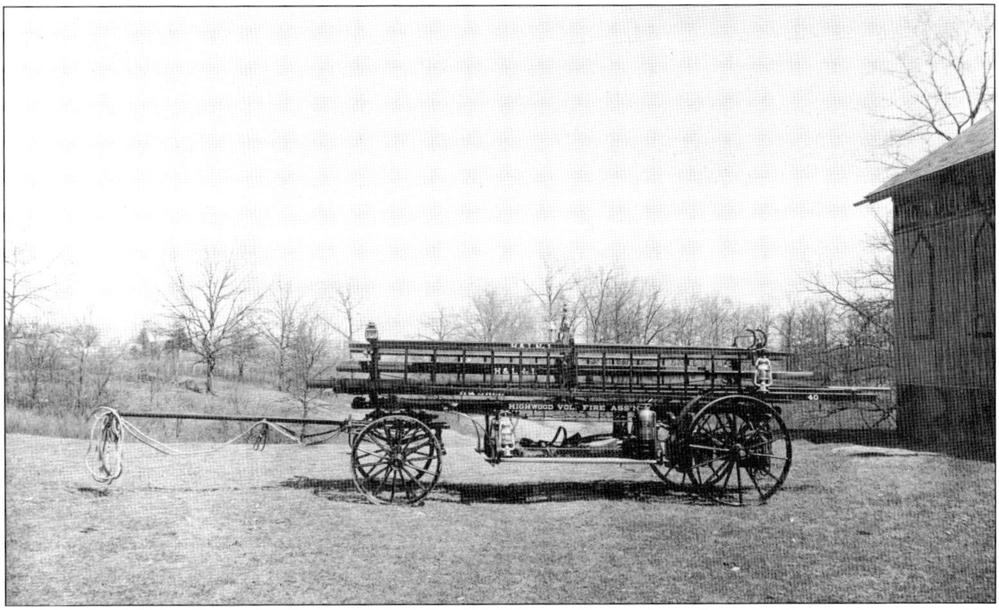

Hamden never had horse-drawn fire apparatus. This hand-drawn ladder truck was constructed by the Highwood volunteers during the winter of 1907–1908. It is pictured here beside their quarters at the New Lebanon Mission. Old-timers told about how the ladder truck was often hitched to the back of a northbound Dixwell trolley car to reach its destination. Hopefully, there were not many stops in between. (HFRA.)

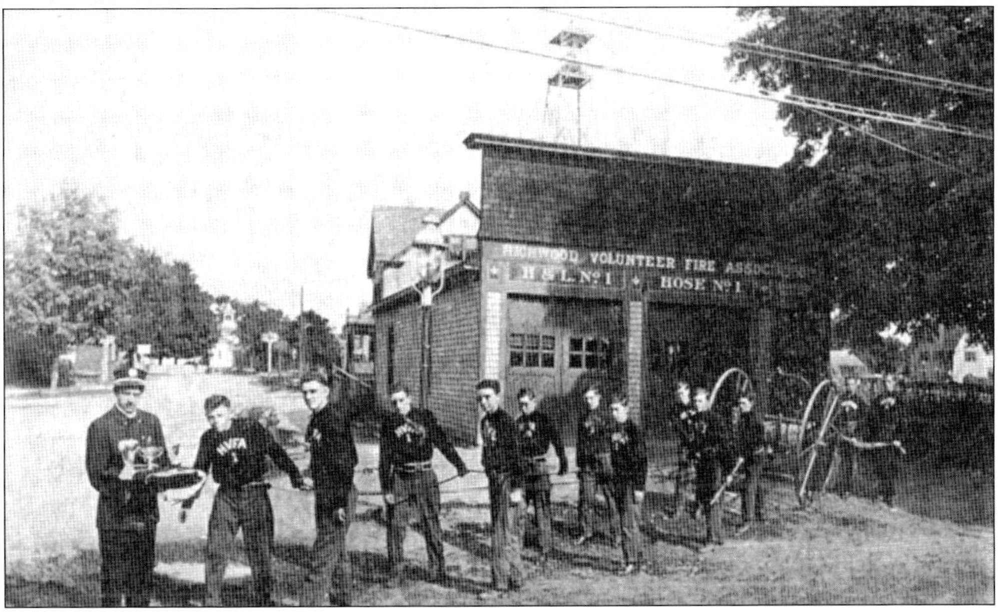

Pictured in front of their new firehouse at Dixwell Avenue and Morse Street, the members of Highwood's 1913 winning hose-laying team are, from left to right, George Visel, William Weise, Lawrence Cashman, Stanley Mautte, Daniel McMullen, George Yardley, Charles Visel, Truman Gilbert, Roland Ruwet, Arthur Norman, George Ruwet, Julius Norman, and Marlin Recor. Later, Roland Ruwet and the Norman brothers all became full-time career members of the department. (HHS.)

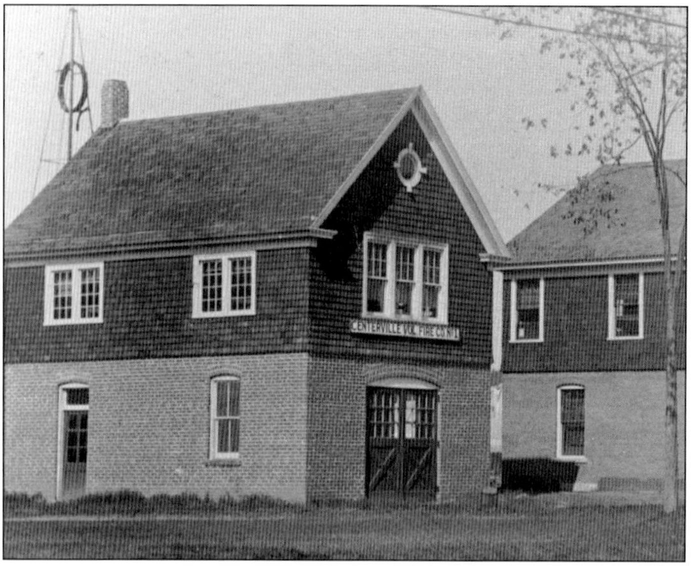

The first meeting of Centerville Volunteer Fire Co. was held in October 1904 at Finley's barn, where the Connecticut Motor Club stands today. Thomas Hartley was elected as the company's first president. The Web Shop donated a hose cart and other equipment. With $945 raised from public subscription, members built their first firehouse (pictured) in 1907 a few feet north of the present town hall station ramp. (Edward Quimby.)

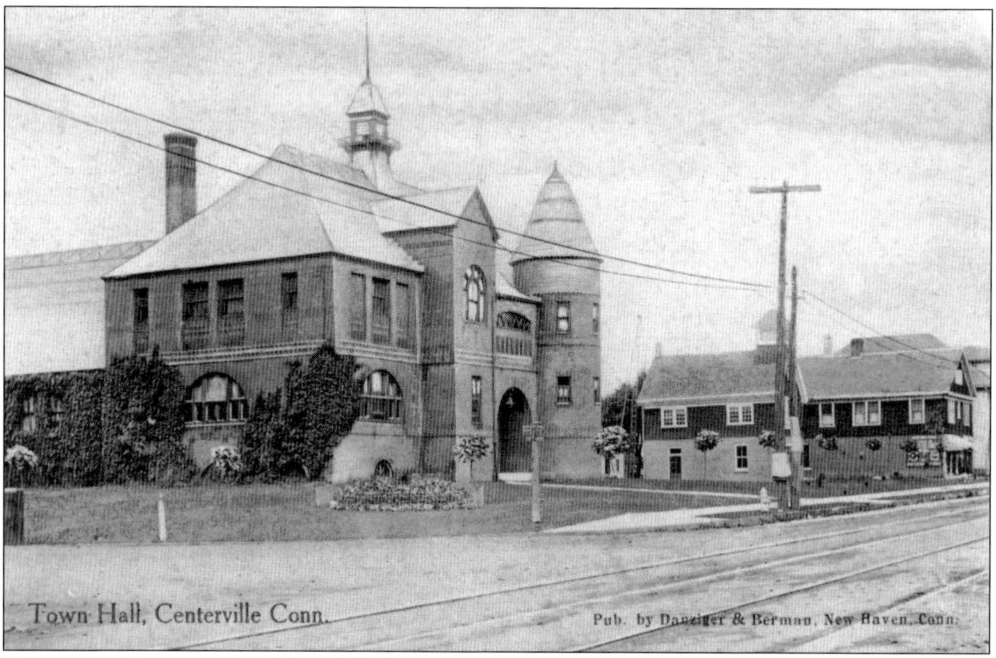

This early-1920s postcard shows the Centerville firehouse in relation to the old town hall. It had gained a cupola belfry since the earlier photograph was taken. Centerville's firehouse and the old town hall were both razed in 1924 for the construction of Hamden's new Memorial Town Hall. The building immediately north of the old firehouse was razed in 2013 for an annex to Eli's-on-Whitney restaurant. (HHS.)

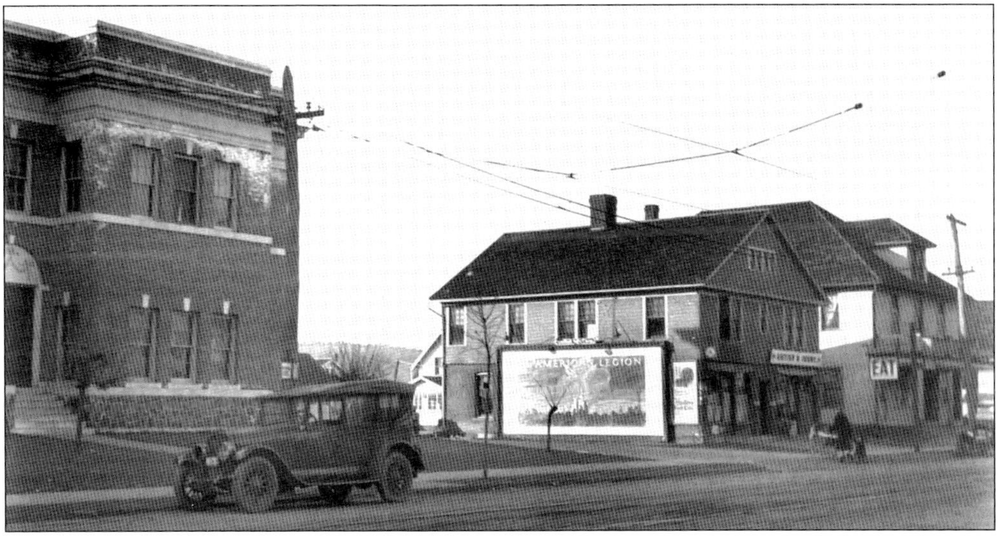

When Hamden's Memorial Town Hall (left) was built in 1924, a new Centerville fire station was incorporated into the portion of the building pictured in this photograph. The apparatus bay doors originally faced north, requiring exiting apparatus to make a 90-degree right turn to enter Whitney Avenue. A 1939–1940 renovation moved the bay doors to their present location at the front of the building. (Wayne Chorney.)

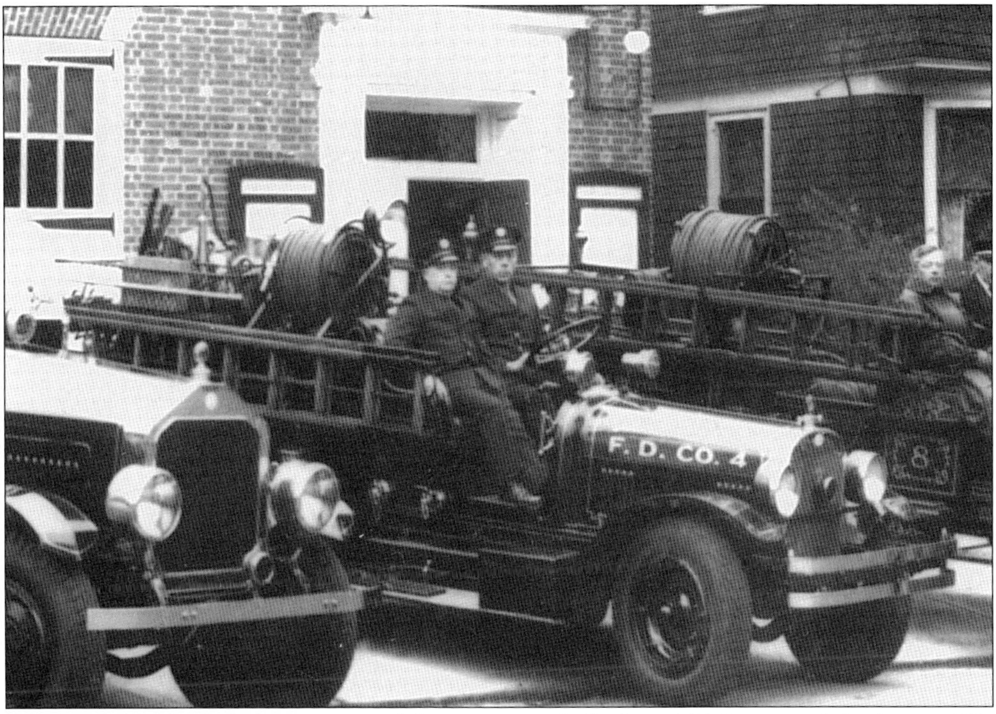

Delivered shortly after the opening of the new town hall, this 1925 Seagrave Suburbanite 350-gpm pumper was Centerville's first motorized apparatus. Pictured in this 1939 photograph at Whitneyville Station 3 are Firefighters Clement Kammerer (left) and Robert Reutenauer. The 1925 Seagrave remained part of the department inventory until the late 1950s, when it was Engine 9 in West Woods. (Chandler Brainard.)

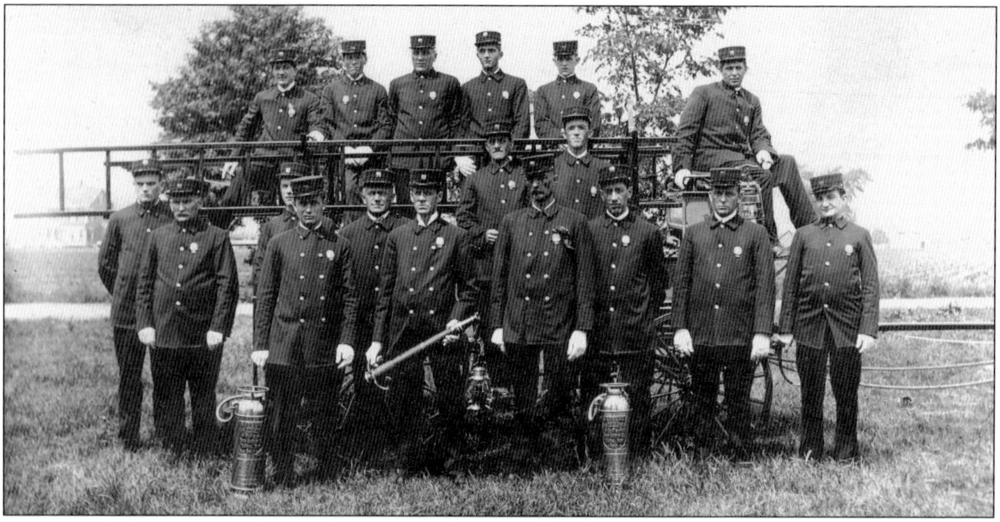

The Humphrey Hook and Ladder Association in the Hamden Plains section grew out of the Humphrey Athletic Club, formed in 1906. The fire company first met in a barn on Dixwell Avenue and was chartered on April 2, 1908. The association's name was derived from the "Humphrey flats" that were spread out over Circular Avenue in the early 20th century. (Joseph Taylor.)

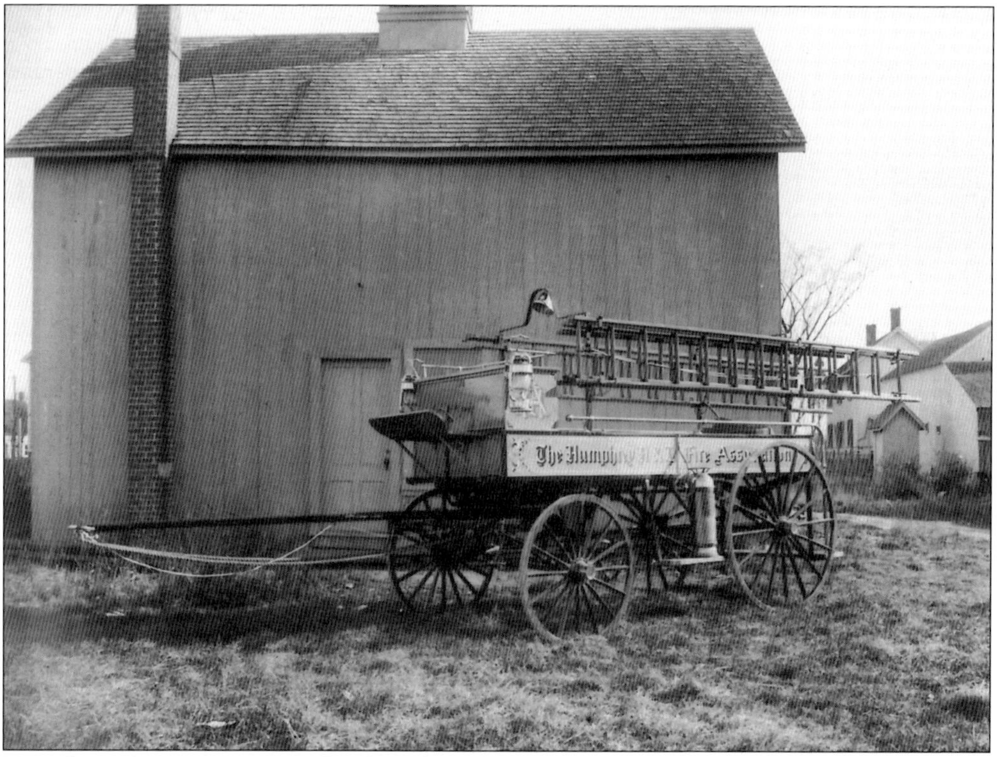

Humphrey's hose wagon was ordered per the July 10, 1908, meeting: "Mr. [Samuel] Flight reported the order for the wagon had been placed with Mr. Lewis. Wagon to be gray with gold lettering, 2 brass lanterns, 1-24 ft., 2-12 ft., and 1 9 ft. ladders, brass railing run all around the wagon. $75 was ordered paid as 25% of payment of wagon. Wagon delivered on Nov. 27, 1908." (Humphrey Hook and Ladder Association.)

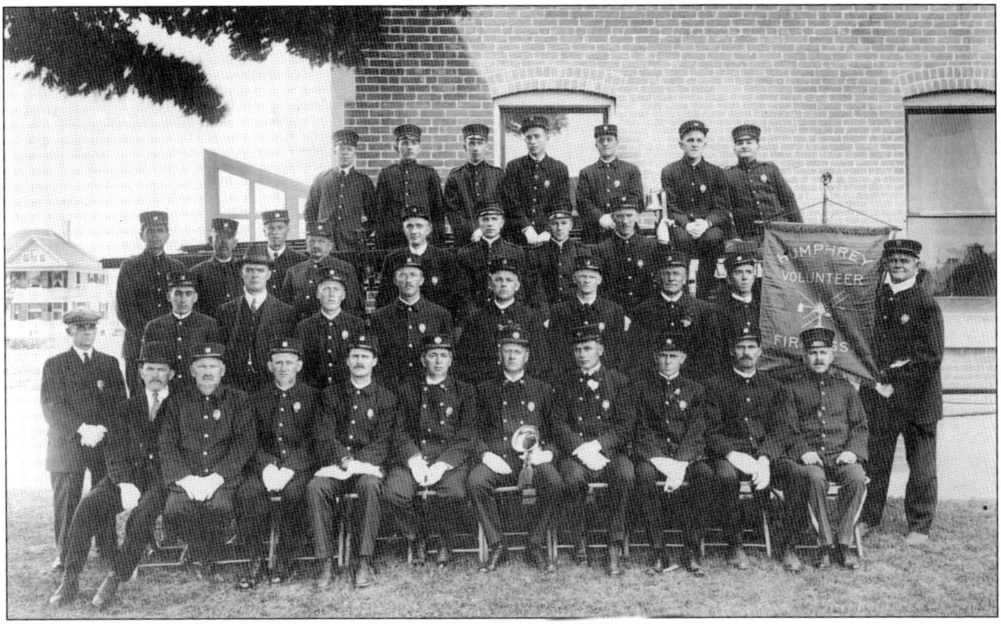

In 1911, land for a new fire station was donated by prominent businessman Virginius Mayo. The Humphrey fire station was dedicated on January 3, 1913. In this photograph taken about 1915, the members of the fire company are assembled beside the fire station as if ready for a parade. Paid fire personnel have been stationed here continuously since 1918. (HHS.)

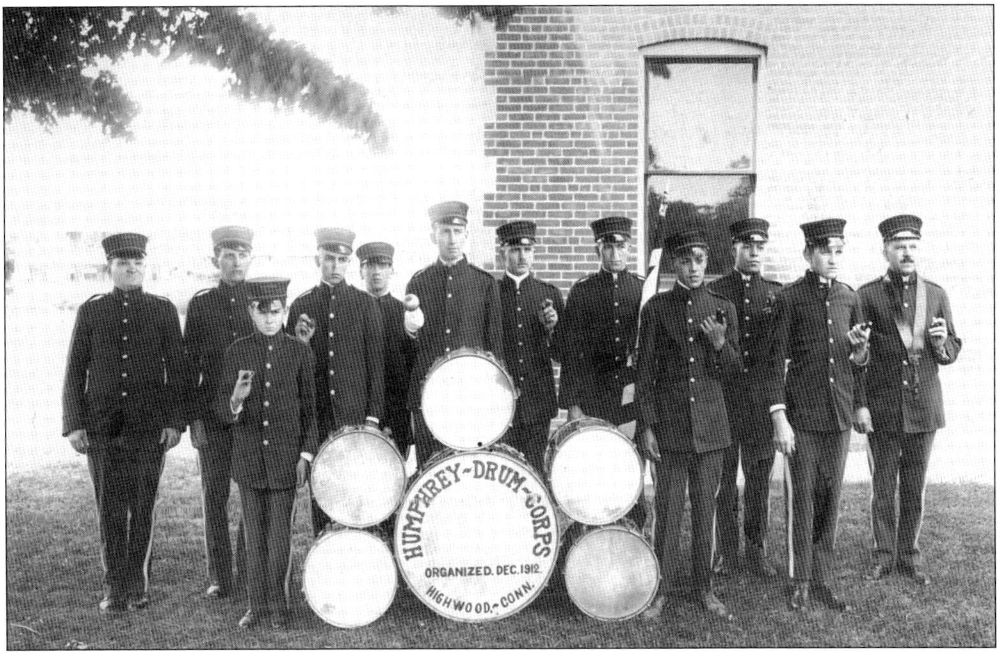

The fire company sponsored the Humphrey Drum Corps, which was very active in area fife and drum competitions during the 1910s. The members in this 1915 photograph are, from left to right, Charles Lennox, Charles Smith, Carlton Crateau, Fred Sturtze, Pie Ornley, James McTiernan, Meyer Hansen, Emil Olson, Emil Sturtz, Oscar Olson, Marshall Murphy, and John Murphy. (Humphrey Hook and Ladder Association.)

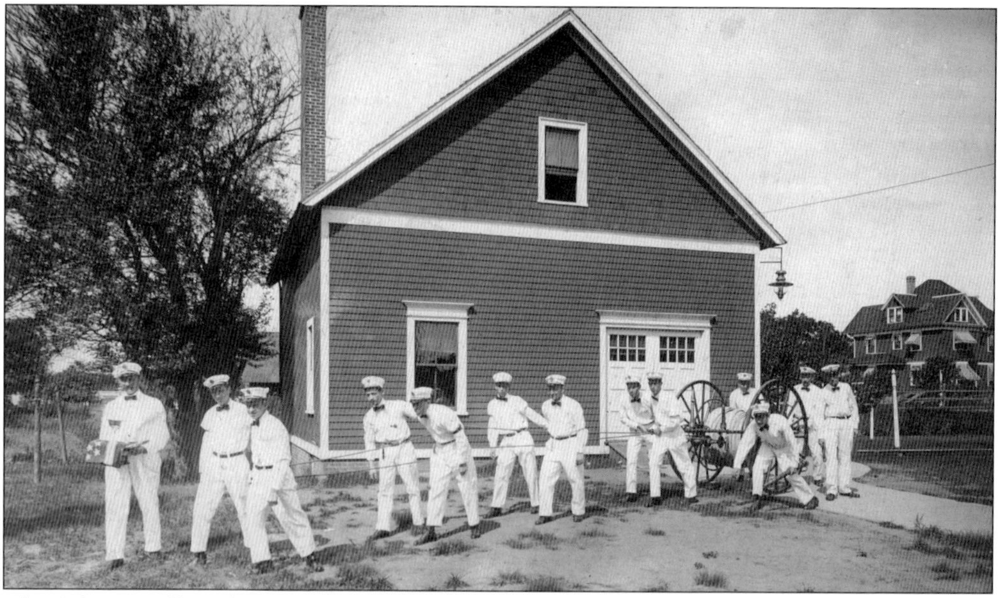

The Whitneyville Volunteer Fire Association was organized in the spring of 1910 in a barn at Whitney and Putnam Avenues. Officially chartered on December 2, 1910, the fire company acquired land at 39 Putnam Avenue. The Winchester Repeating Arms Co. donated a small wood-frame building to serve as a firehouse. The Whitneyville Improvement Society donated a hose cart and 350 feet of hose. (Arthur J. Tefft.)

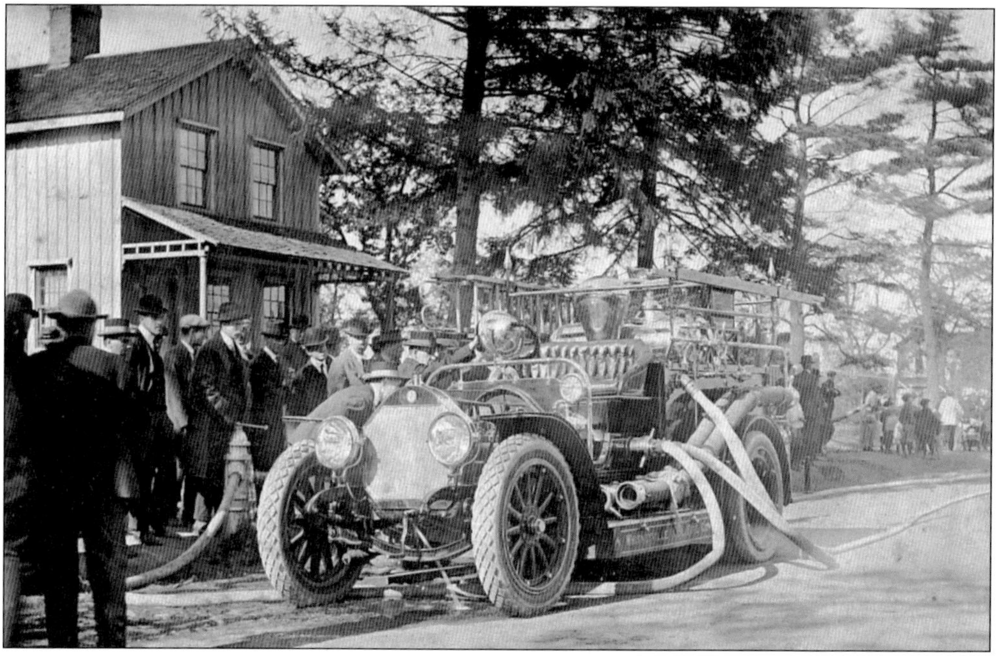

The Whitneyville volunteers were the first to acquire a motorized piece of fire apparatus. After a subscription campaign raised more than enough money for a chemical truck, the town contributed another $1,000 to purchase the $5,500 pumping engine. Delivered in 1915, this Maxim triple combination pumper built on a Thomas Flyer chassis was the first motorized pumping engine manufactured by Maxim. (Arthur J. Tefft.)

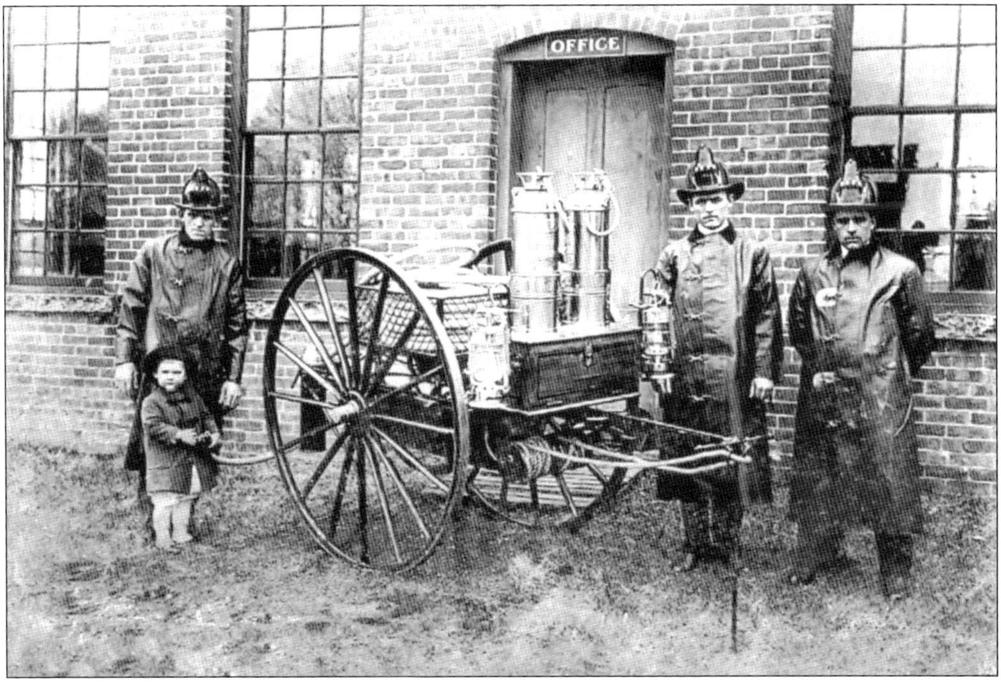

The organizational meeting of the Mt. Carmel Volunteer Fire Co. was held on November 8, 1911, in St. Mary's Hall (below), site of the present Our Lady of Mount Carmel Church. Former first selectman A.E. Woodruff was elected the company's first president and James H. Strain its first captain. Pictured above are, from left to right, Capt. James H. Strain, Bud Fitch, and Arthur Pelton. The small boy is the captain's son, James W. Strain, who would become company captain in the late 1930s and, later, a career firefighter, retiring in 1973 as a deputy chief. The Wirt & Knox hose cart, 500 feet of hose, chemicals, and extinguishers cost $693.30. The W.W. Woodruff & Sons Co. building in the background, at Broadway and Ives Street, was destroyed by fire on January 12, 1968. (Above, Strain family; below, HHS.)

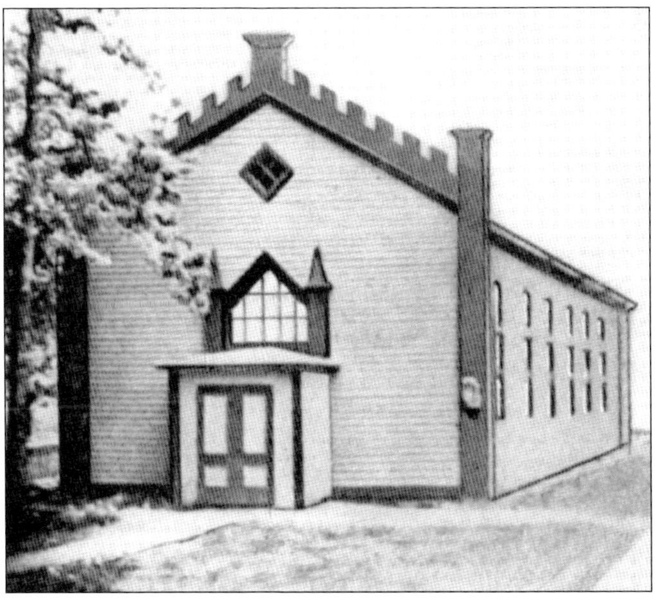

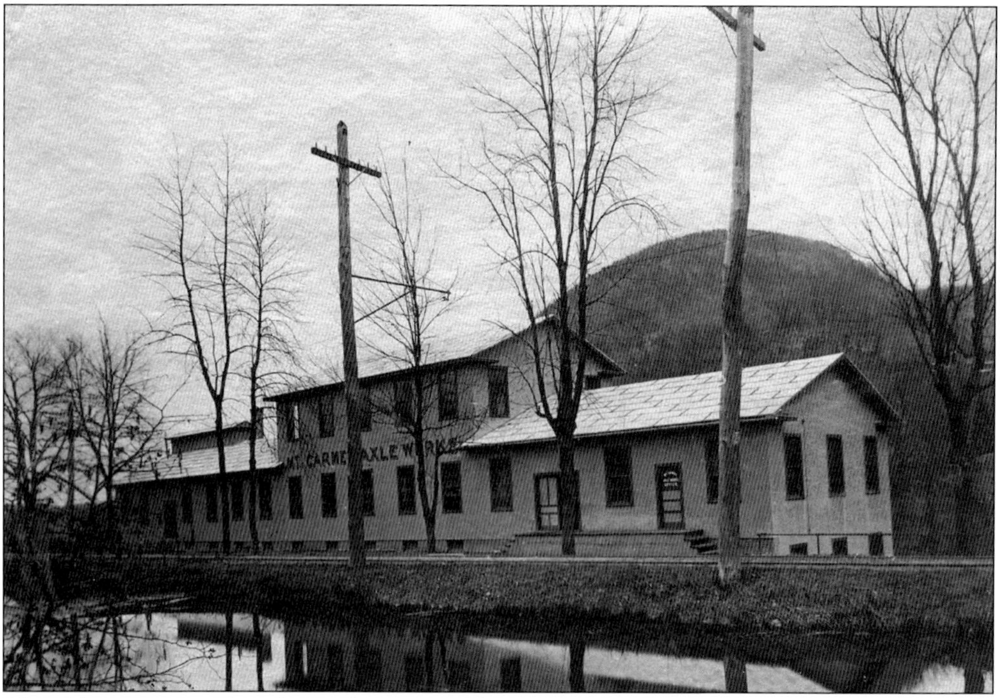

The old Farmington Canal is in this view of the "new" Mount Carmel (Upper) Axle Works on Whitney Avenue in what was called Mount Carmel Center. It replaced the original upper axle works building that burned down in January 1905. The North Hamden Improvement Society's volunteer fire company was organized here in December 1911, when it was the Liberty Cartridge Company. The volunteers disbanded in 1926. (HHS.)

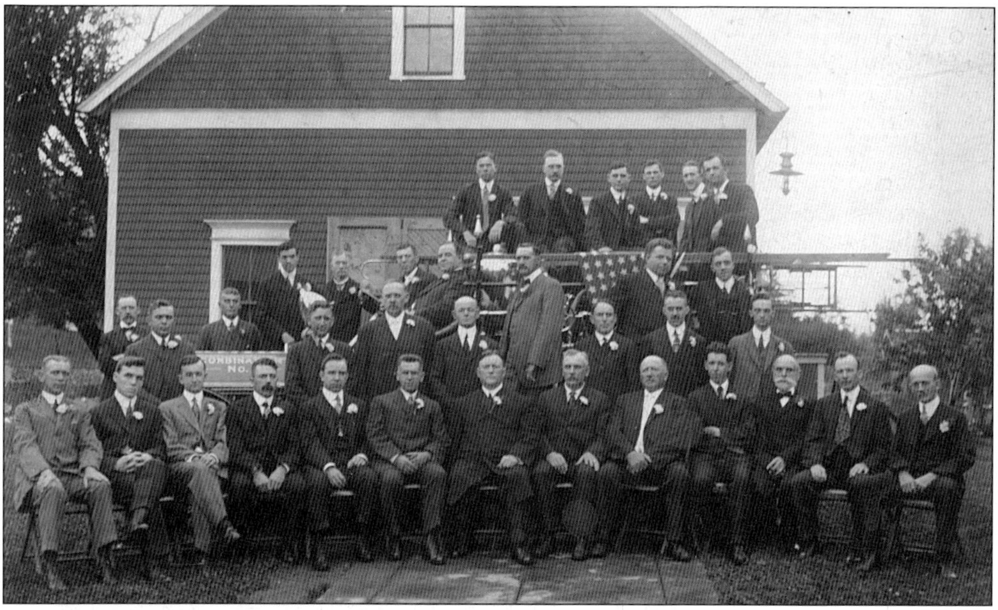

The delivery of Whitneyville's new Maxim pumper was a major community event. The entire Whitneyville fire company is pictured here on June 21, 1915, with its new "machine" in front of its quarters at 39 Putnam Avenue. (Arthur J. Tefft.)

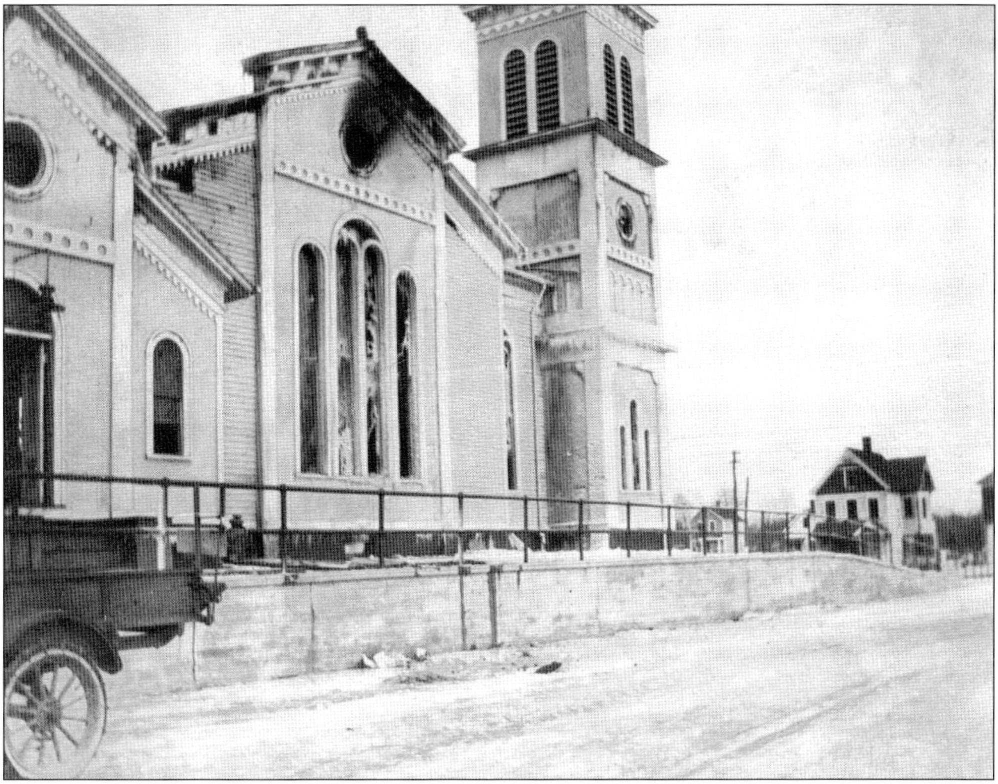

On Sunday, January 13, 1918, a faulty chimney flue caused a fire at the Hamden Plains Methodist Church, the first Hamden fire for which there are known photographs. The Humphrey, Highwood, and Whitneyville volunteers and a motorized New Haven fire company fought the blaze in bitter cold temperatures. This photograph of the aftermath shows the heavy damage to the 85-year-old church, which was a total loss. (HHS.)

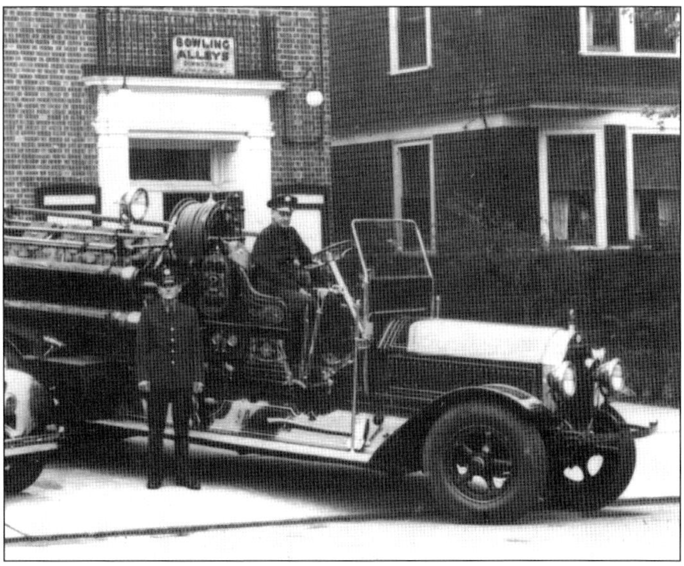

The 1918 church fire led to motorization of three more fire companies. Before the year was out, the Highwood and Mount Carmel companies each acquired Stewart chemical trucks. The town also appropriated $11,039 to purchase this Seagrave 750-gpm pumper for the Humphrey volunteers in 1919. Firefighters Joseph Hromadka (left) and Edward Kromer are pictured with this pumper in front of Whitneyville Station 3 in 1939. (Chandler Brainard.)

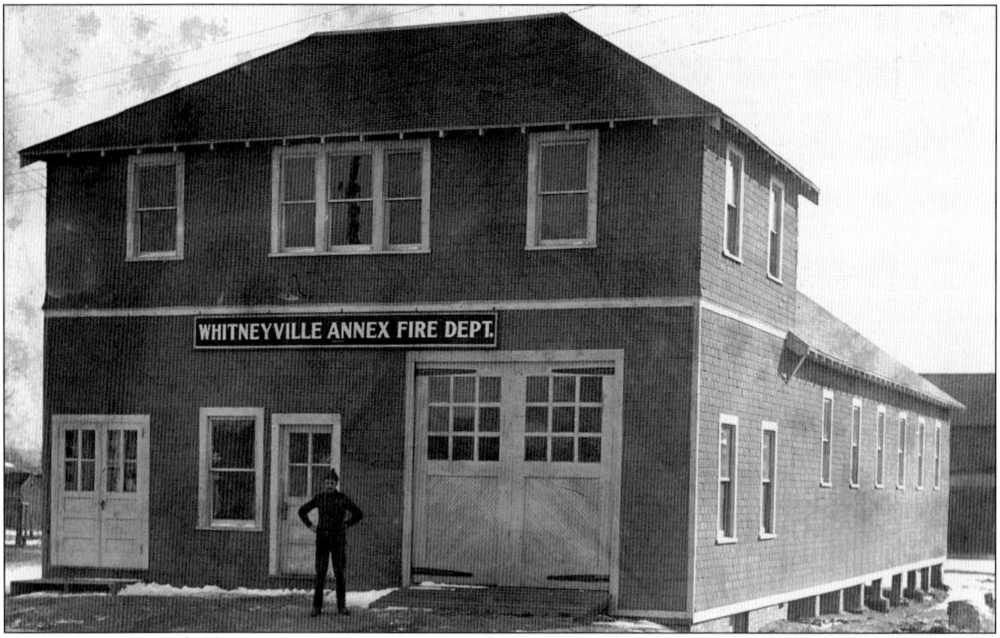

In April 1921, the Whitneyville volunteers sponsored the formation of the Whitneyville Annex Volunteer Fire Department to provide fire protection to the southeastern part of town along State Street. The members built their new station at the corner of Merritt and Webb Streets. Following the 1925 formation of the Hamden Fire Department, the company was renamed Merritt Street Volunteer Fire Co. No. 6. (HHS.)

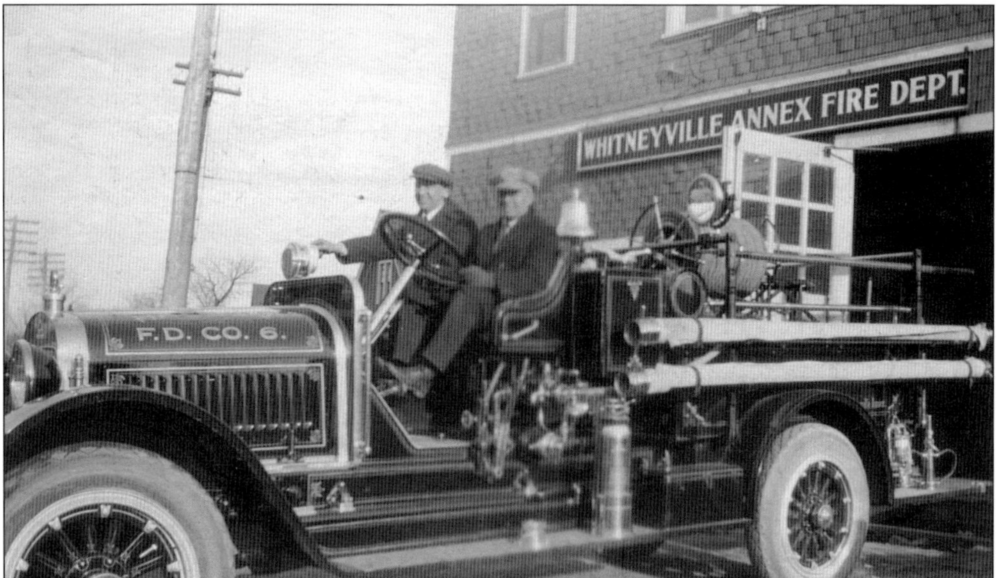

The town allocated $7,150 to purchase this Stutz 350-gpm pumper, seen here when delivered to Whitneyville Annex in April 1924. The driver is unidentified. August Schoepfer (left) is operating the hand-cranked siren. The pumper remained in service until April 1942. It was eventually sold to Co. 7 member Dwight Kirk, who added a 400-gallon tank. The pumper was sometimes pressed into service when needed. (HHS.)

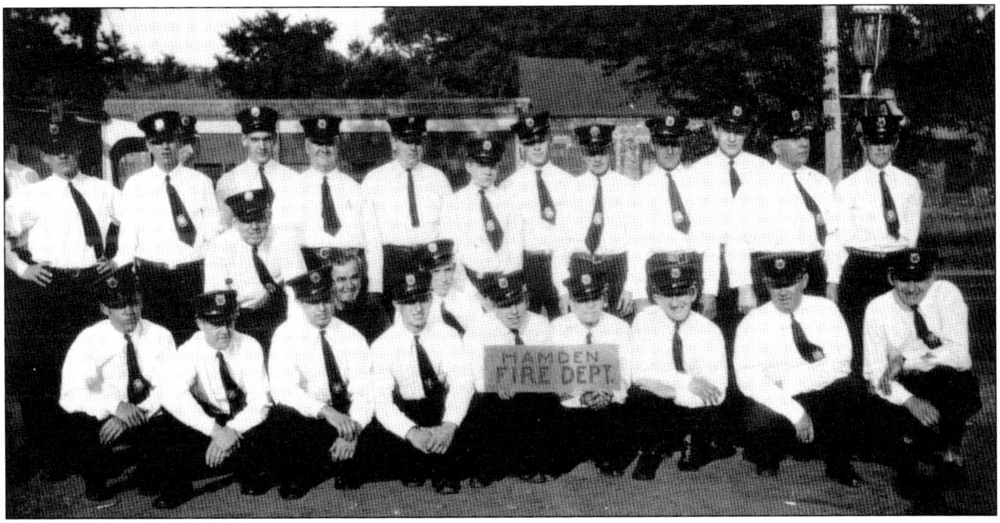

The members of the Mt. Carmel Volunteer Fire Co. are pictured beside their firehouse before leaving for the 1929 Connecticut State Firemen's Convention and parade at East Hartford on August 17. Raymond K. Spencer (kneeling far right) would be president of the company from the 1950s until the late 1980s. (Gilbert Spencer.)

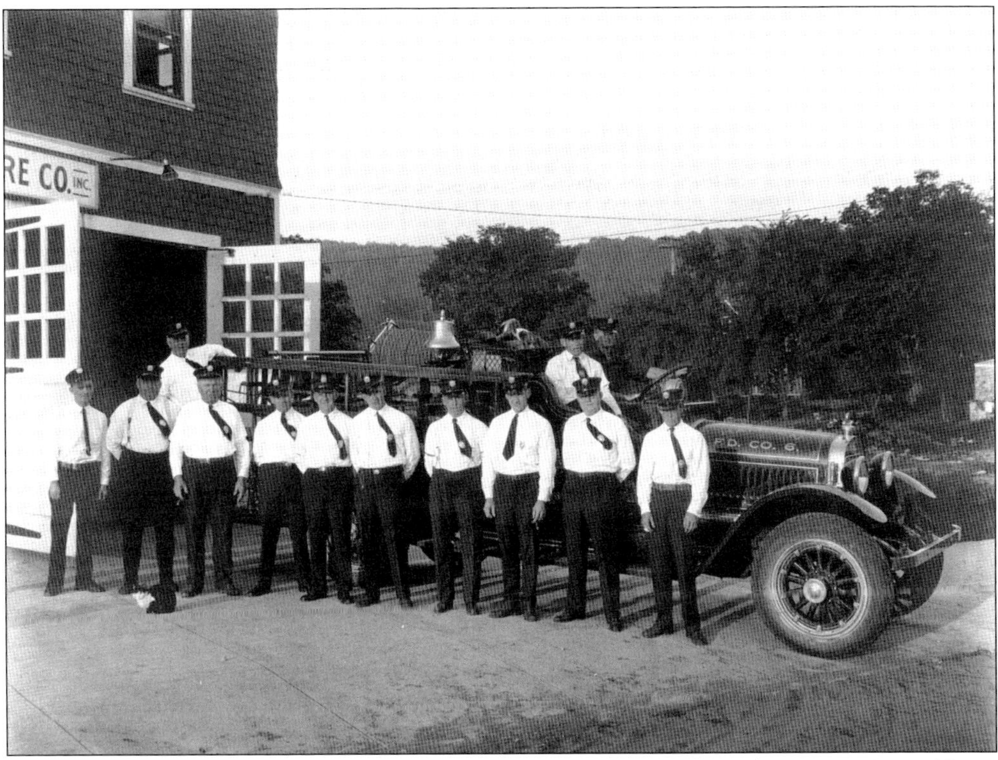

The Merritt Street volunteers may be preparing for the same parade as the Mount Carmel boys, as they pose beside their 1924 Stutz pumper. Standing are, from left to right, Michael Parillo, James Clerkin, Oliver Abed (on tailboard), Michael Lipinsky, James Minahan, August Schoepfer, Lawrence Spahr, John Schoepfer, Michael Durso, William Meldrum, and John Kimler. Capt. Harry Stein is sitting next to paid driver George Benton. (Arthur J. Tefft.)

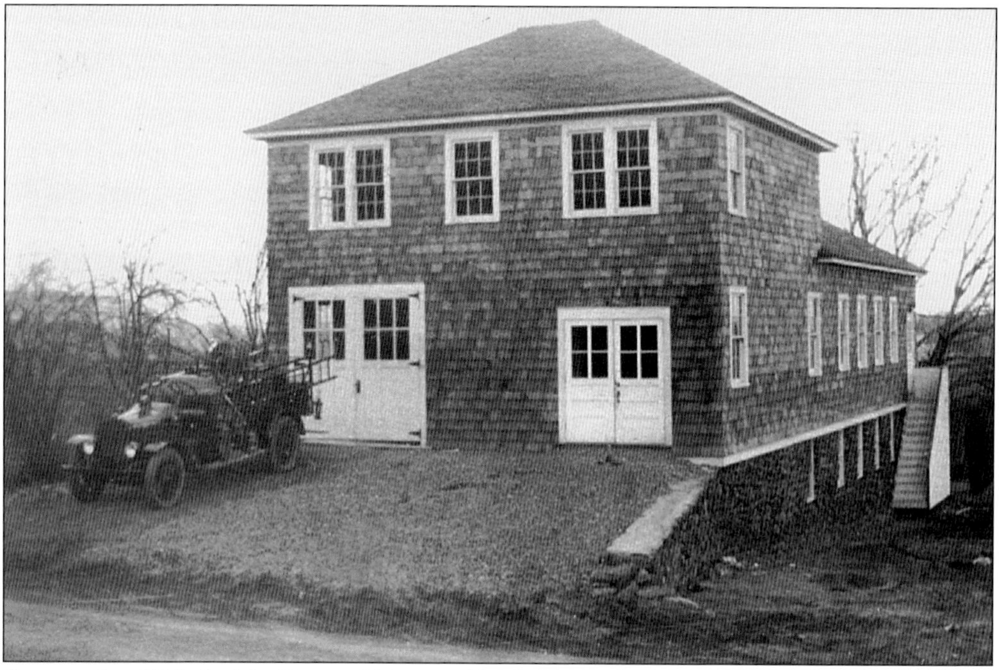

Organized in 1924, the Mix District Volunteer Fire Co. No. 7 dedicated its new Shepard Avenue fire station in April 1928. Parked on the ramp is the Stewart chemical truck that was originally purchased for Highwood in 1918. This volunteer fire company is still active in 2017. (Joseph Taylor.)

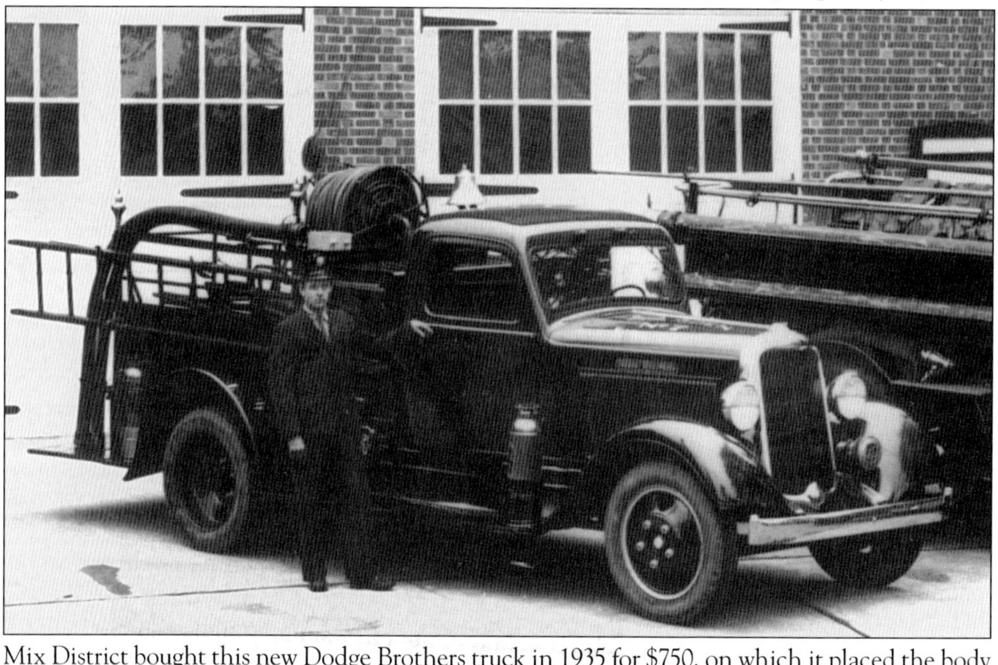

Mix District bought this new Dodge Brothers truck in 1935 for $750, on which it placed the body of the 1918 Stewart. Charter member Clement Wetmore is pictured beside the truck in this May 1939 photograph taken at Whitneyville Station 3. He would serve as superintendent of alarms and apparatus until his 1968 retirement. Wetmore's son Paul would serve as chief from 1993 to 1996. (Chandler Brainard.)

Two

The Hamden Fire Department

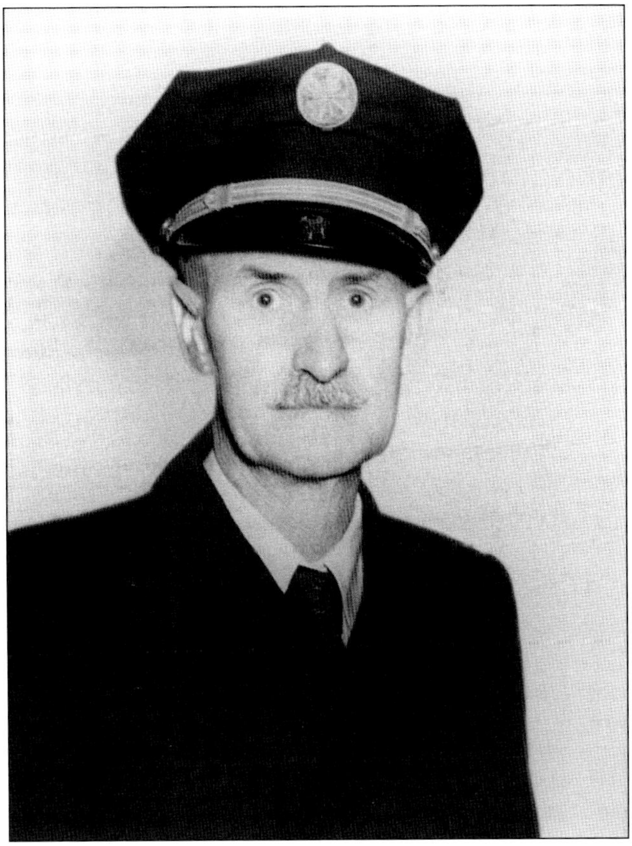

A 1925 act of the state legislature created the Hamden Board of Fire Commissioners, which appointed Charles Loller as Hamden's first fire chief. Chief Loller, also the town's building inspector, had been captain of the Highwood volunteers since 1900. The paid drivers and volunteer firefighters all worked under the direction of Chief Loller, his assistant chiefs, and the elected line officers of the volunteer companies. (Fire Chief's Office.)

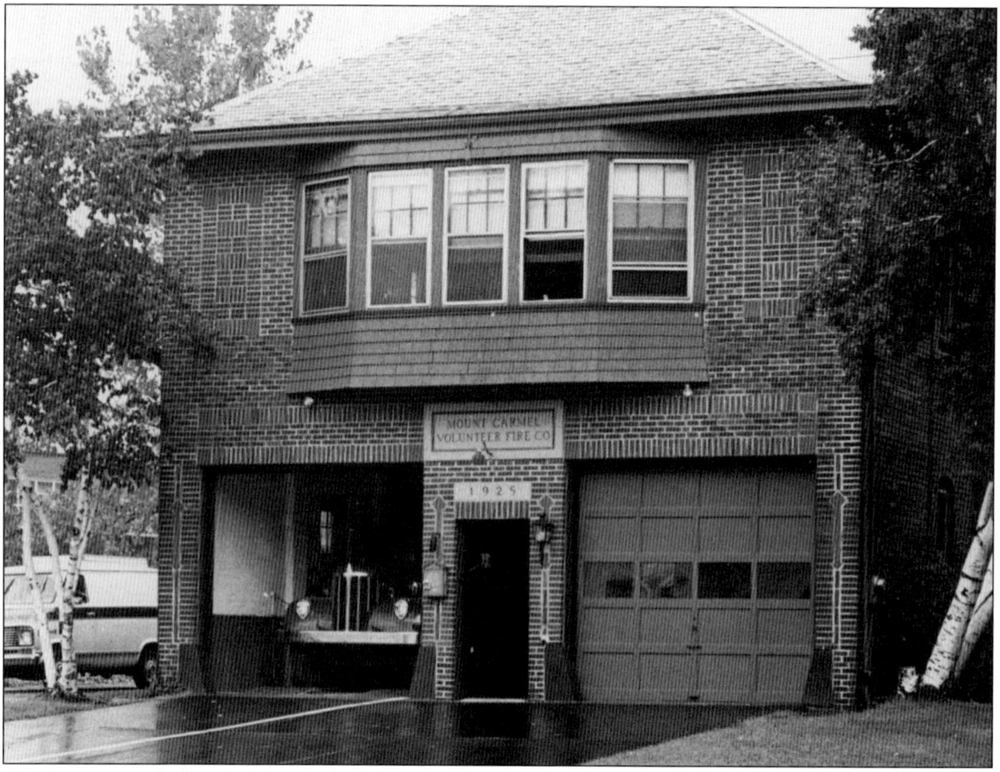

In August 1925, the Mt. Carmel volunteers bought land at 2993 Whitney Avenue on which to build their firehouse, which opened in January 1926. On October 25, 1926, the members elected Albert Purce their paid driver. Career personnel have been stationed at the Mt. Carmel station ever since. This photograph was taken in 1974, the year before the volunteers built an annex to the station. (Kenneth DeLorme.)

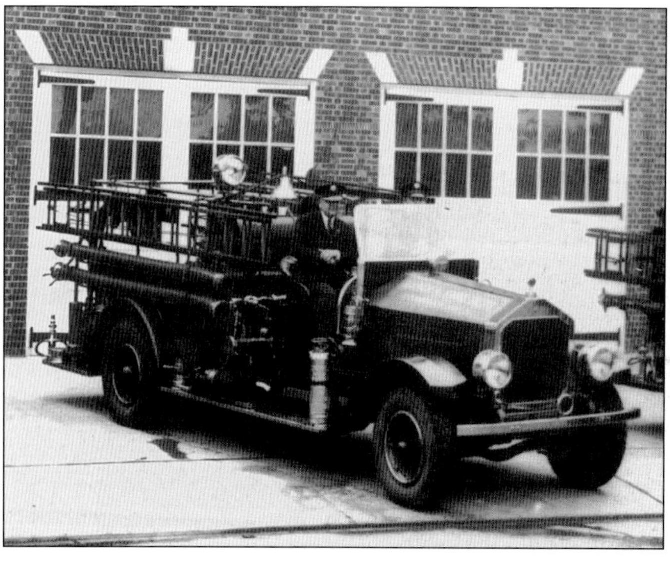

On January 6, 1930, this Maxim 600-gpm pumper was delivered to Mt. Carmel, where it served until it was transferred to Merritt Street in 1953. Hamden was a good customer of the Middleboro, Massachusetts, manufacturer, purchasing a total of twelve pumpers and three ladder trucks from Maxim over the course of 58 years. The rider in this 1939 photograph at Whitneyville Station 3 is Albert Molleur. (Chandler Brainard.)

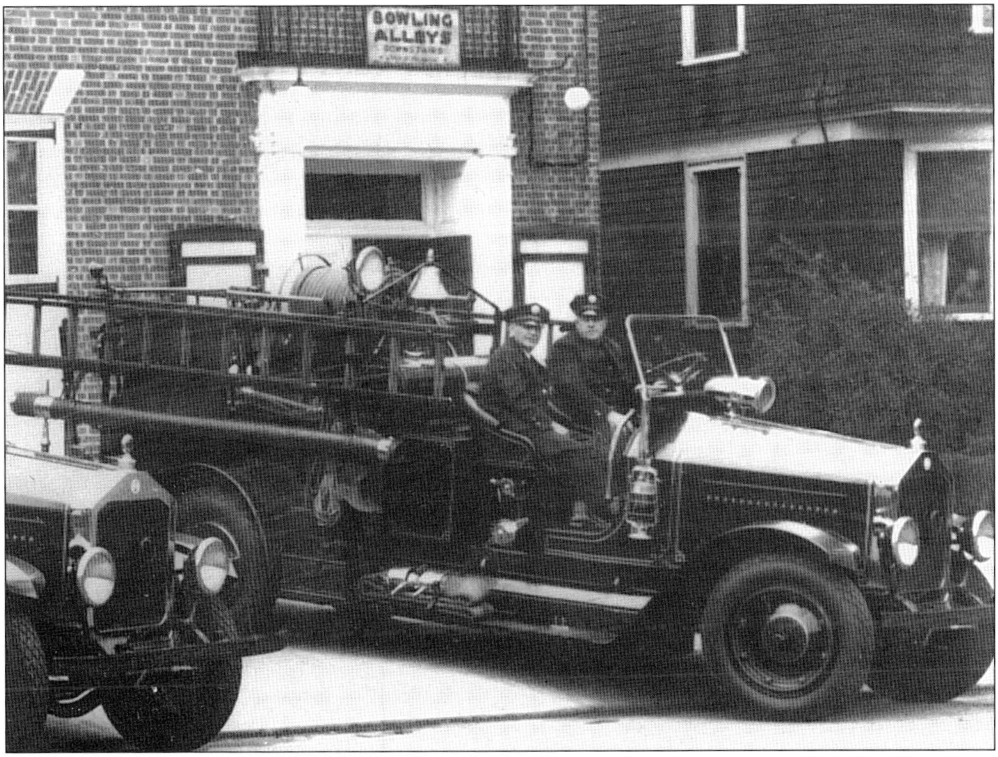

This Maxim 500-gpm pumper, delivered to Highwood in March 1926, was part of a $12,500 package deal that included the ladder truck pictured below. Pictured above in May 1939 at Whitneyville Station 3 are Firefighter Joseph Marchitto (left), also known as "Joe Marks," and the driver, Firefighter Joseph Dukat. Below is a rare factory photograph of Hamden's first motorized ladder truck, delivered to Highwood in March 1926 with the 500-gpm pumper pictured above. Along with a full complement of ground ladders, the truck was equipped with a chemical tank and booster line. In March 1941, the truck was struck by a trolley car at Dixwell Avenue and Mather Street and declared a total loss. (Above, Chandler Brainard; below, Matthew Lee.)

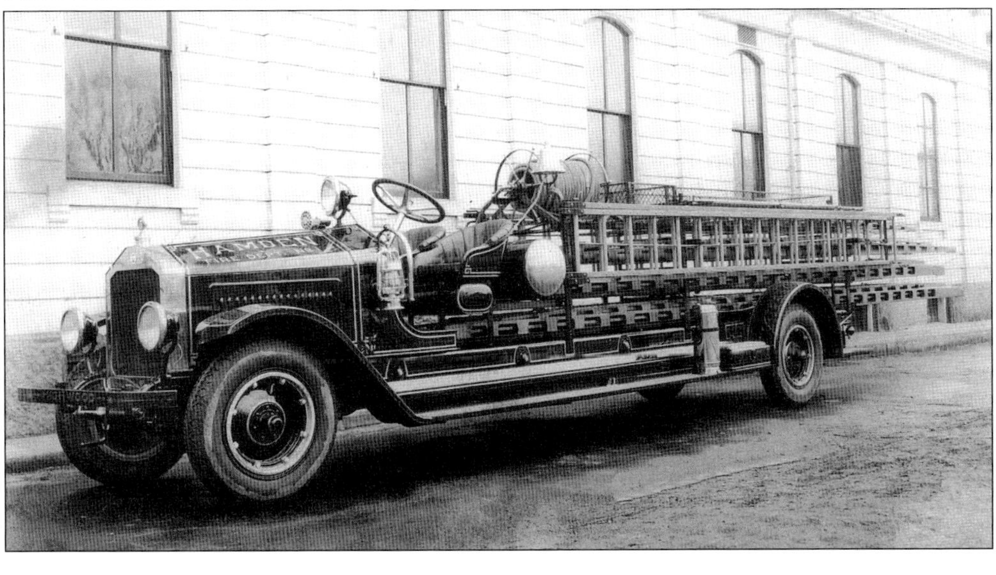

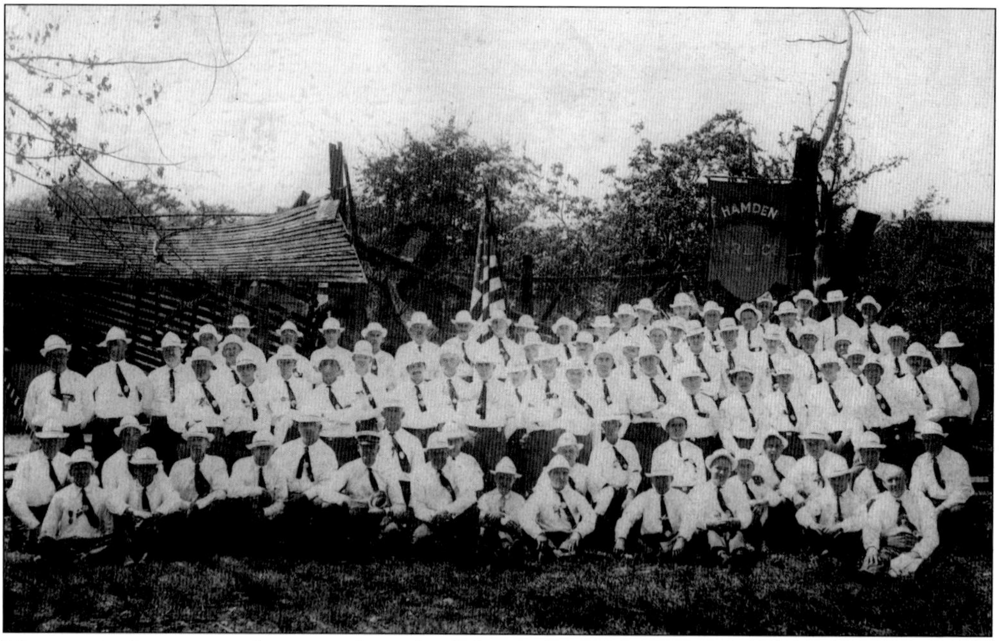

The esprit de corps and camaraderie firefighters demonstrate on the fireground are just as evident when marching in parades or competing in the field days that have been major firefighter extracurricular activities since the earliest days. On August 20, 1927, at West Haven's Savin Rock, Hamden's volunteer fire companies united to form the largest marching contingent, for which they won a trophy. (HHS.)

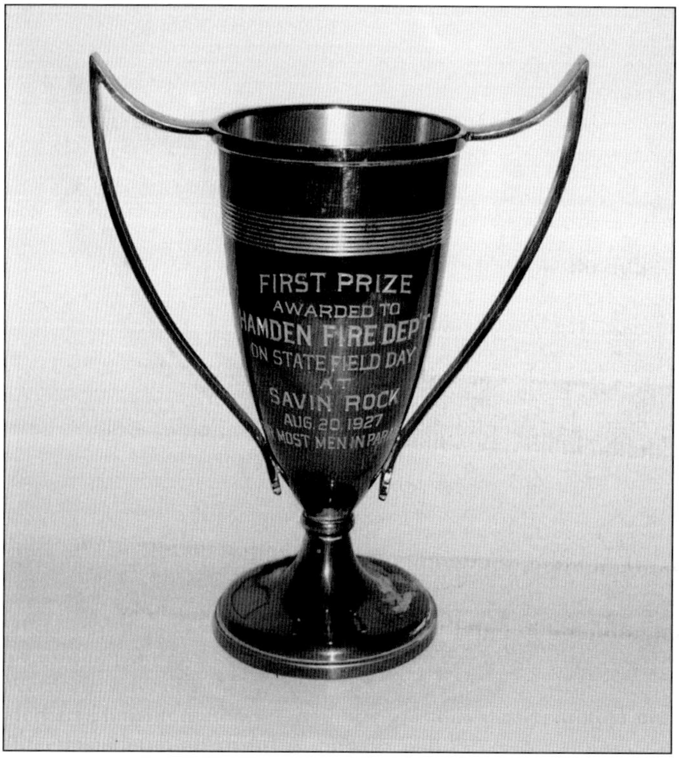

This impressive trophy was the pride of the department when Hamden's volunteer firefighters returned from the Connecticut State Field Day at Savin Rock in August 1927. Savin Rock was a popular amusement park in West Haven from the late 19th century until it closed in 1966. (HFRA.)

On Saturday, December 10, 1927, Centerville Volunteer Firefighter Edward Meegan and six other firefighters responded to an alarm in Highwood on Centerville's 1925 Seagrave. On Whitney Avenue near Servoss Street, the pumper's rear wheel hit a raised trolley track, flipping the truck over. Four of the firefighters were unhurt. Two received only minor injuries. Firefighter Meegan was seriously injured and died the next day. (*New Haven Evening Register.*)

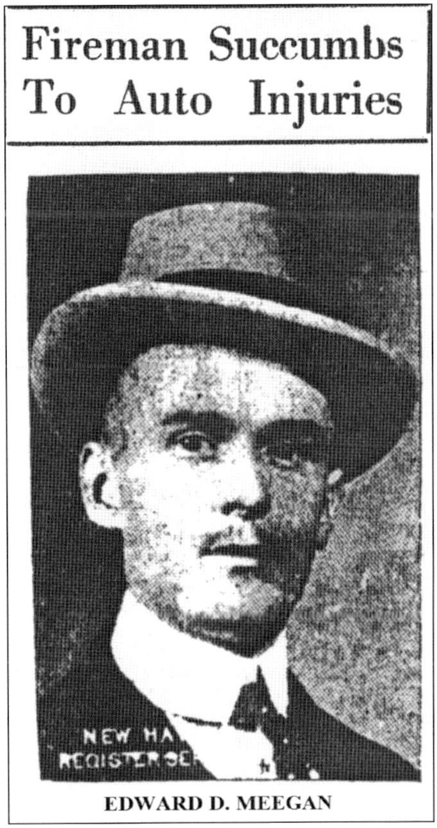
Fireman Succumbs To Auto Injuries

EDWARD D. MEEGAN

Dunbar Hill Volunteer Fire Co. No. 8 was organized in May 1926. This photograph, taken outside the station at Dunbar Hill Road and Benham Street, is believed to have been about the time Co. 8 members had acquired Whitneyville's 1910 Locomobile in 1928. The two gentlemen wearing the helmets are brothers Nelson (left), the company's first vice president, and Everett Warner, the company's first captain. (HHS.)

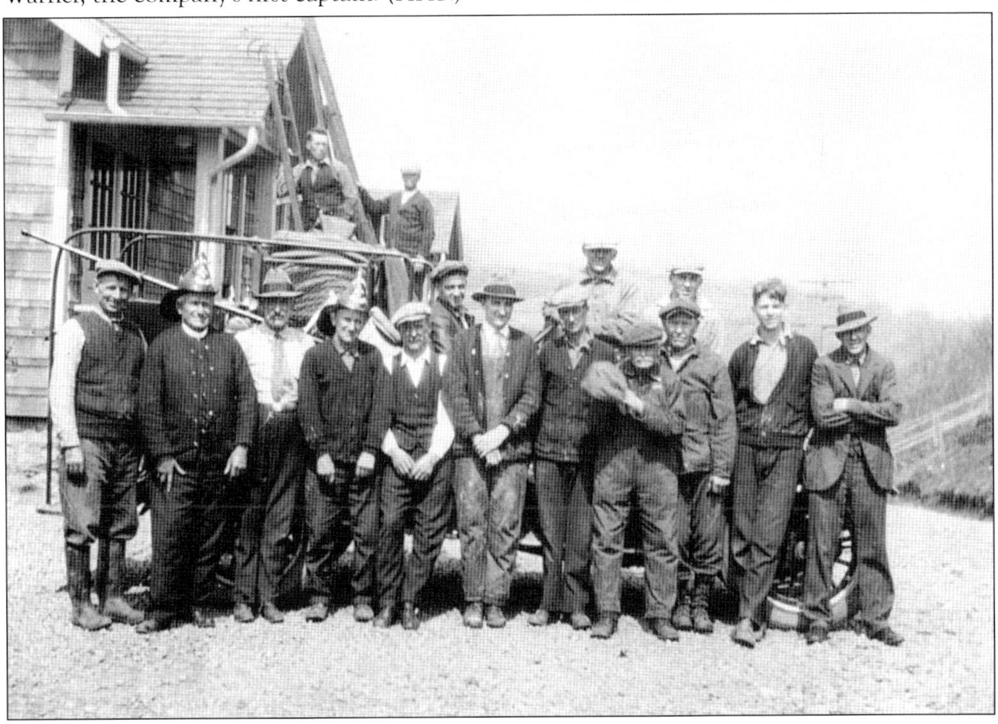

HAMDEN FIRE DEPARTMENT

Fire Chief, Charles P. Loller, 956 Dixwell Ave. Colony 6256
Assistant Chiefs { C. Raymond Brock, 21 Augur St. Pioneer 5893-J
Raymond C. Spencer, 2545 Whitney Ave. Ham. 88 }

FIRE COMMISSIONERS

Edgar W. Munson, 1270 Dixwell Avenue. Colony 264
William Balke, 59 Putnam Avenue. Colony 2641-3
Thomas Miller, 2911 Whitney Avenue. Hamden 1229
Clerk—W. H. Avis, 154 Putnam Ave. Colony 2405

No. 1—Highwood
Colony 2503
913 Dixwell Avenue

President......M. J. Whalen
Recording Secretary....David Howe
Financial Secretary..Julius Norman
Treasurer......Frank Loller
Drivers...... { Joseph Marks
Arthur Norman }

Hose Company

Captain......Geo. O'Connell
First Lieutenant......M. J. Whalen
Second Lieutenant......Edw. Welch

Hook and Ladder

Captain......Julius Norman
First Lieutenant......Arthur Norman
Second Lieutenant......David Howe

No. 2—Humphrey
Colony 2502-2
71 Circular Avenue

President......George W. Warner
Captain......Walter Kling
First Lieutenant......F. Edmonds
Second Lieutenant.....R. MacKenzie
Secretary......R. Eno
Drivers...... { Charles Coe
Leslie Coe }

No. 3—Whitneyville
Colony 2501-2
37 Putnam Avenue

President......Henry F. Hall
Vice-President......William Balke
Finan. Sec. & Treas. James Bulger
Recording Secretary....Fred Morton
Captain......Chas. Guptil
First Lieutenant......J. Endicott
Second Lieutenant.....Ralph Rosson
Drivers...... { Francis Nolan
Earl Gussman }

No. 4—Centerville
Hamden 124-2
Memorial Town Hall

President......Franklin Spencer
Secretary......John Scanlon
Captain......Frank Allaire
First Lieutenant......Edward Crook
Driver......John Vroman

No. 5—Mt. Carmel
Hamden 65
2995 Whitney Avenue

President......John Whitham
Vice-President......Joseph Judge
Treasurer......William Trappe
Secretary......Frank Hatch
Captain......James Strain
First Lieutenant......Henry Adams
Second Lieutenant..Walter Shelomis
Driver......Albert Purce

No. 6—Merritt Street
Liberty 3980
Merritt corner Webb Street

President......Chas. O. Atwater
Captain......Henry J. Stein
First Lieutenant......Wm. Meldrum
Second Lieutenant......M. Dertzo
Secretary......C. M. Simpson
Driver......Robert Reutenauer

No. 7—Mix Department
Hamden 243
Shepard Avenue

President......David Cooper
Vice-President......R. Morse
Captain......Dwight Kirk
Secretary......C. Wetmore
Treasurer......W. W. Morse
First Lieutenant......W. Wetmore

No. 8—Dunbar Company
Hamden 259-4
Dunbar Hill

President......W. H. Hindinger
Vice-President......Nelson Warner
Record. Secretary..Harold Emerson
Captain......Everett Warner
First Lieutenant......Lewis Bradley

This 1927 Hamden city directory entry is perhaps the earliest fire department roster of the volunteer fire company officers and paid drivers. More paid firefighters would be added in the next few years, but the volunteer chiefs and duly elected volunteer captains and lieutenants were in charge at emergencies until the department was reorganized in April 1942. (HHS.)

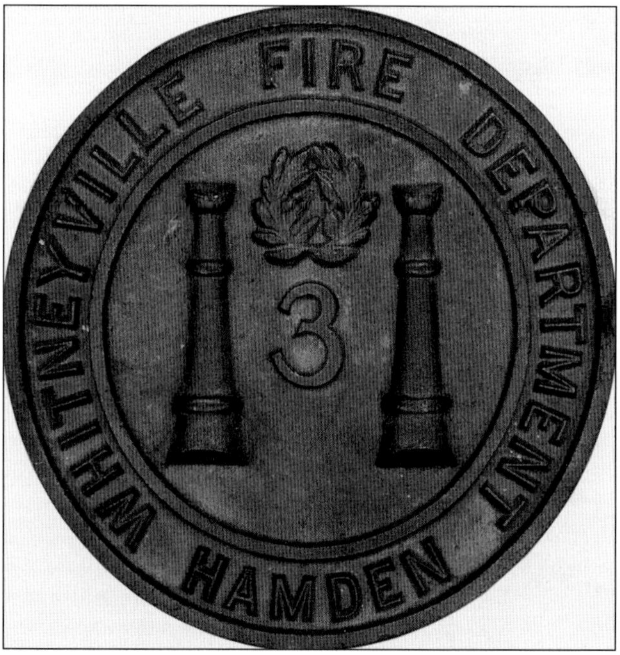

Grave markers for deceased Hamden firefighters, like this one for a Whitneyville member, can be found in several Hamden cemeteries. Most may be found in the Whitneyville cemetery at the corner of Whitney Avenue and Treadwell Street. (Elizabeth Onofrio Smith.)

These gentlemen were the leaders of the department in the mid-1930s. From left to right are (first row, seated) Commissioners Thomas Miller, James Gillies, and Albert Ruwet, Chief Charles Loller, and Assistant Fire Marshal Thomas Pender; (second row, standing) Assistant Chiefs August Ball, Thayer Jones, and Raymond C. Spencer, and an unidentified gentleman. (HFRA.)

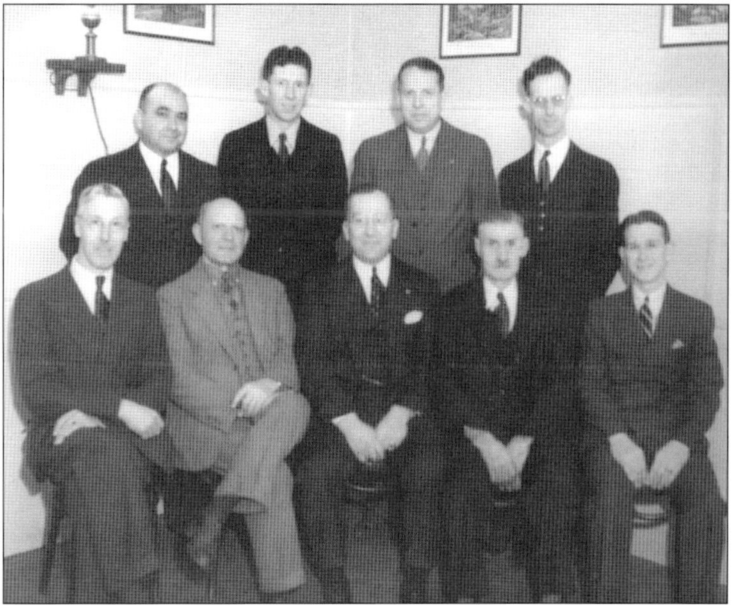

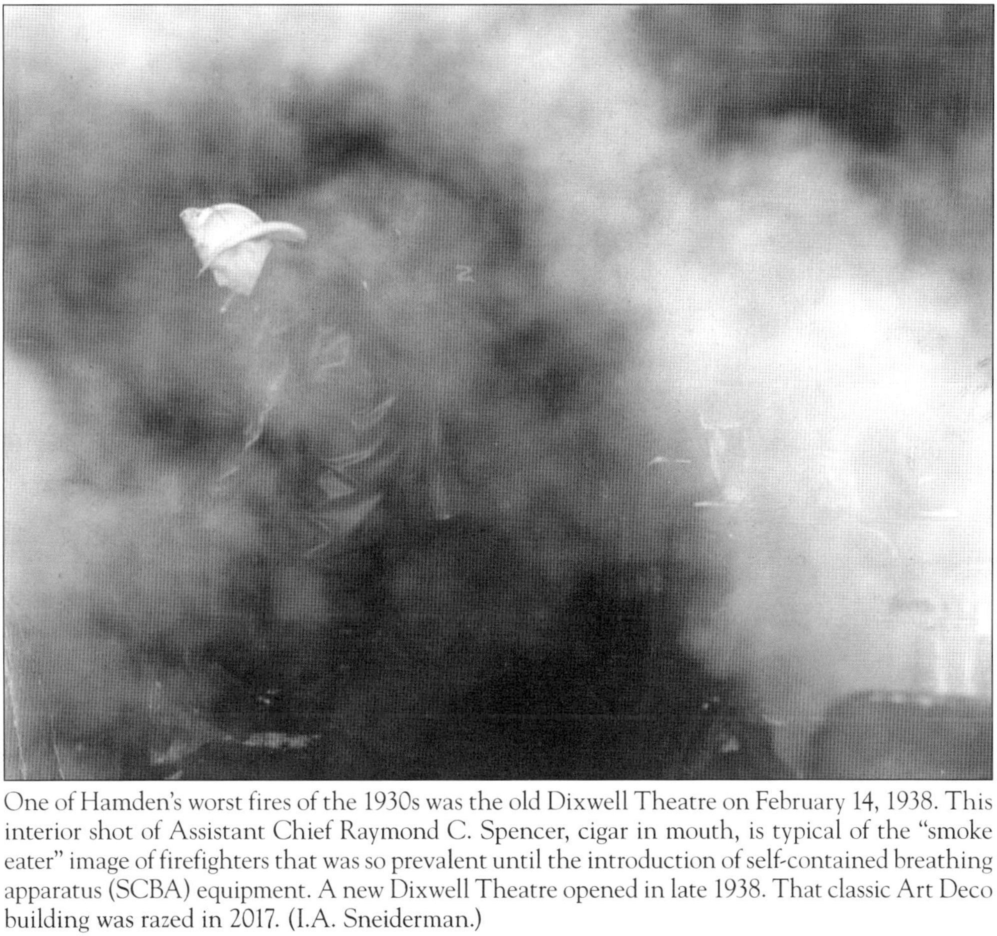

One of Hamden's worst fires of the 1930s was the old Dixwell Theatre on February 14, 1938. This interior shot of Assistant Chief Raymond C. Spencer, cigar in mouth, is typical of the "smoke eater" image of firefighters that was so prevalent until the introduction of self-contained breathing apparatus (SCBA) equipment. A new Dixwell Theatre opened in late 1938. That classic Art Deco building was razed in 2017. (I.A. Sneiderman.)

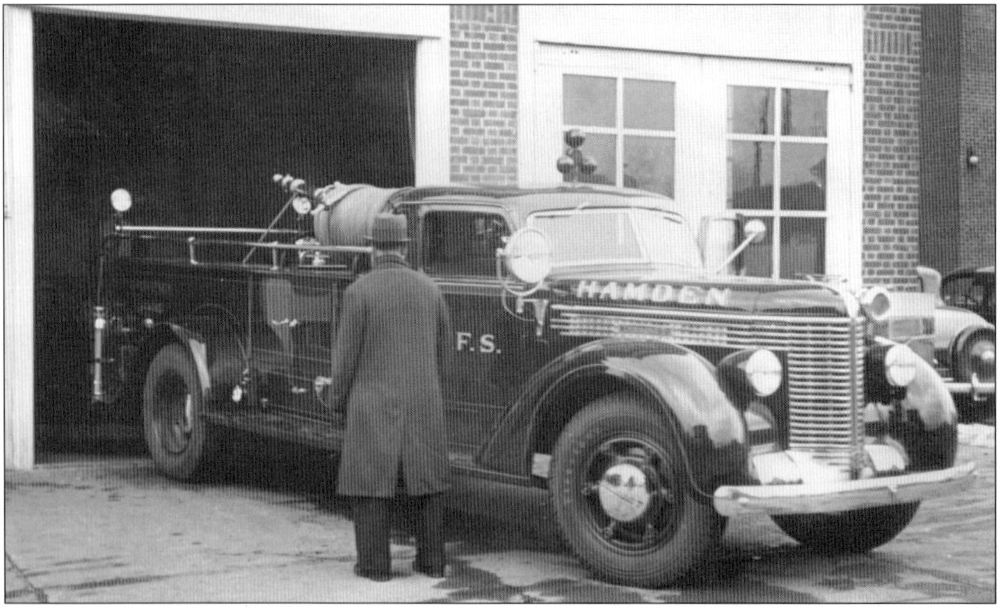

This page features two views of the north side of Centerville Station 4, before the bay doors were relocated to face Whitney Avenue. This 1938 Diamond-T Emergency Squad was the first of five Hamden apparatus manufactured by the Wood Engineering Services of Topsfield, Massachusetts. The Squad responded with the officer on virtually every call. It became volunteer Engine 9 in 1959 and was sold in 1971. (G. Donald Steele.)

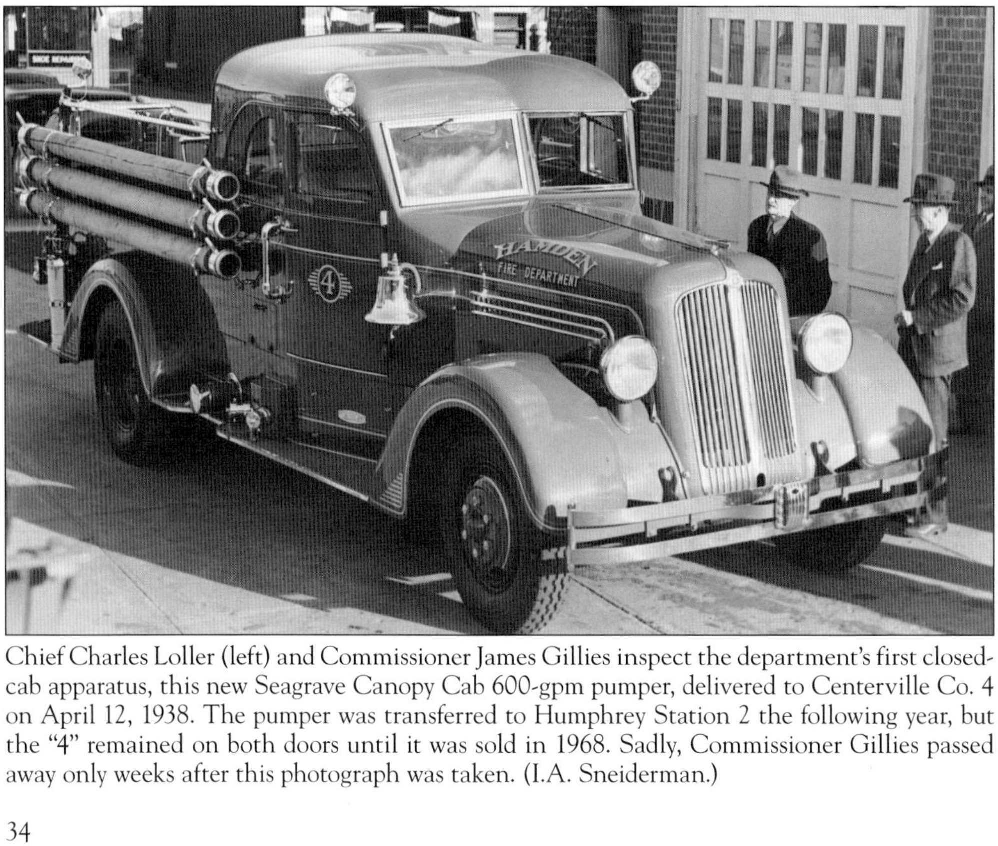

Chief Charles Loller (left) and Commissioner James Gillies inspect the department's first closed-cab apparatus, this new Seagrave Canopy Cab 600-gpm pumper, delivered to Centerville Co. 4 on April 12, 1938. The pumper was transferred to Humphrey Station 2 the following year, but the "4" remained on both doors until it was sold in 1968. Sadly, Commissioner Gillies passed away only weeks after this photograph was taken. (I.A. Sneiderman.)

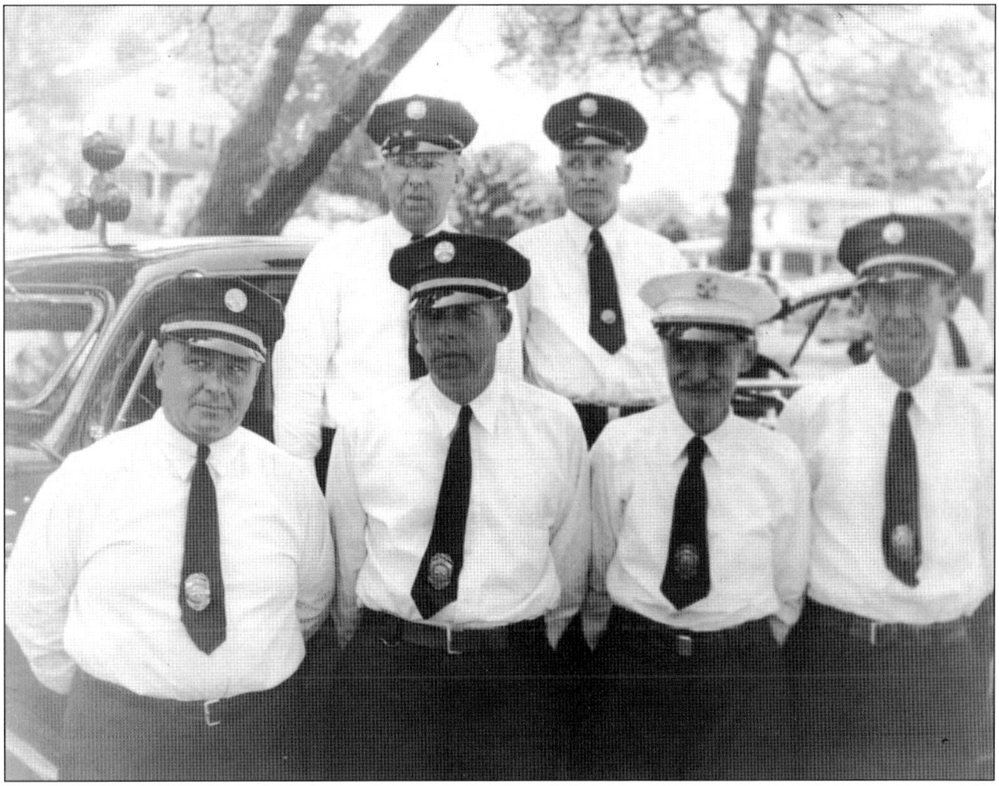

Department brass attended the Connecticut State Firemen's Association Convention parade at Milford in August 1938. From left to right are (first row) Assistant Chiefs August Ball and Raymond C. Spencer, Chief Charles Loller, and Assistant Chief Thayer Jones; (second row) Commissioner Albert Ruwet, Commissioner Albert Ruwet, and Commissioner Leroy L. Jackson. (G. Donald Steele.)

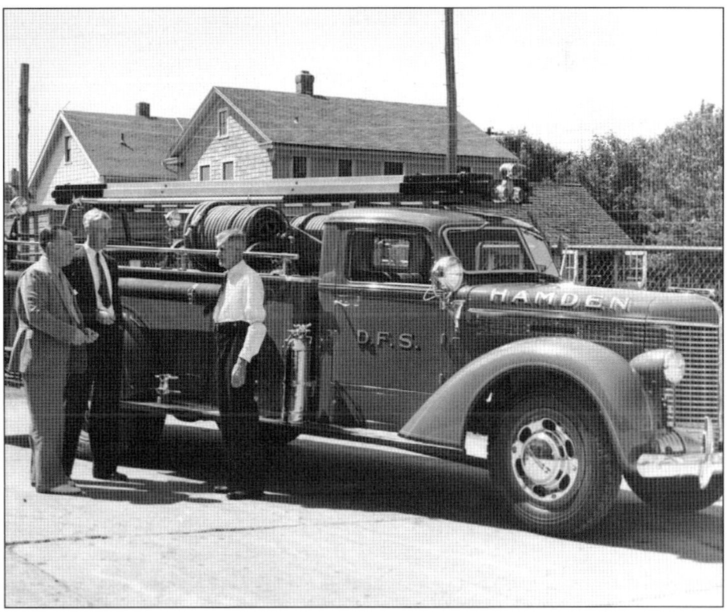

The second delivery from Wood Engineering Services was this 500-gpm Diamond-T pumper, arriving on June 16, 1939. Commissioners Albert Ruwet (left) and Thomas Miller are pictured with Chief Charles Loller as they inspect the new pumper, which was assigned to Centerville Station 4. Hamden's new Seagrave pumper, delivered the previous year, was transferred to Humphrey Station 2. (I.A. Sneiderman.)

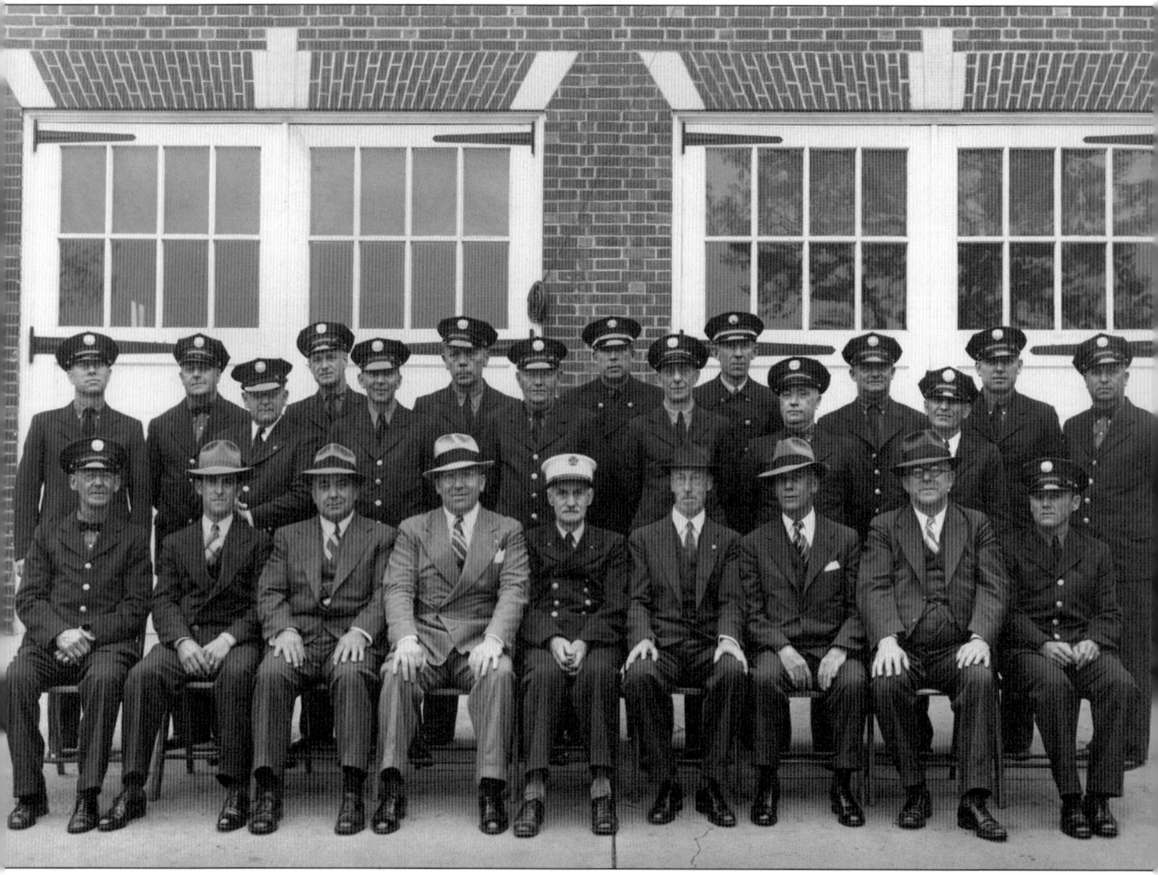

In May 1939, the department leadership and all career personnel posed in front of Whitneyville Station 3 on Putnam Avenue. From left to right are (first row, seated) Firefighter David F. Howe, alarm superintendent Elton Wetmore, Assistant Chief August Ball, Commissioner Albert Ruwet, Chief Charles Loller, Commissioners Thomas Miller and Leroy Jackson, commission clerk Benjamin Bamford, and Firefighter Joseph Hromadka; (second row, standing) Firefighters Roland Ruwet, Arthur Norman, Albert Molleur, Edward Kromer, Francis Nolan, Robert Reutenauer, and Joseph Dukat, First Assistant Chief Raymond C. Spencer, Firefighter Ralph Rosson, Assistant Chief Thayer Jones, and Firefighters Clement Kammerer, Everett Doherty, Joseph Marchitto, Albert Purce, and Julius Norman. This was when the department was still administered by the Hamden Board of Fire Commissioners and Volunteer Fire Chief Charles Loller, who also served as Hamden building inspector. Chief Loller, his assistant chiefs, and the volunteer line officers, duly elected by their respective fire companies, directed all paid and volunteer personnel on the fire ground. (Paul Wetmore Sr.)

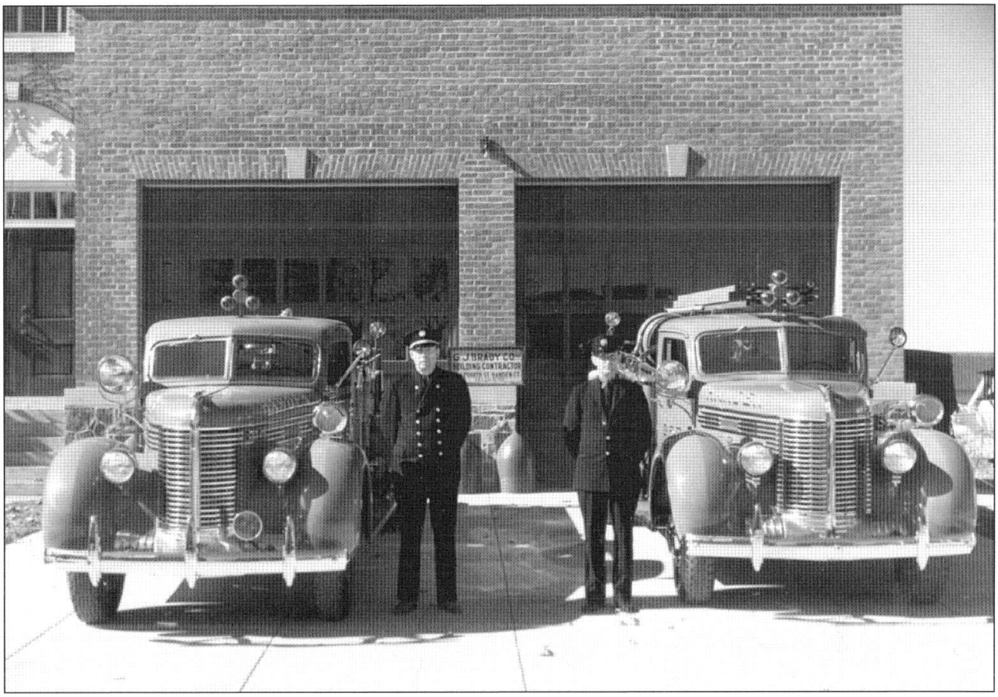

Legendary local photographer I.A. Sneiderman was at Centerville Station 4 to take this photograph after the 1939–1940 renovation that relocated the apparatus bay doors to face Whitney Avenue. Standing by the 1938 Diamond-T Squad (left) and Engine 4, the 1939 Diamond-T pumper, are Assistant Chief Raymond C. Spencer (left) and Firefighter Walter Thomas. (I.A. Sneiderman.)

The "Alarm Room" was added to the back of Centerville Station 4 in 1941. Previously, fire dispatch was handled out of Humphrey Station 2. When Station 4 apparatus were out, telephone and dispatch capabilities were transferred to another station until a cover company arrived at Station 4. The Alarm Room was Hamden's fire dispatch center until Central Communications opened when 911 was introduced in November 1981. (HFRA.)

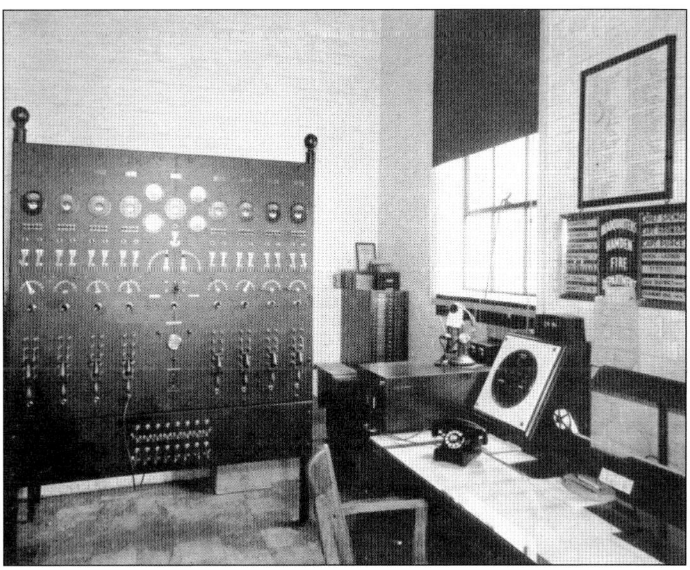

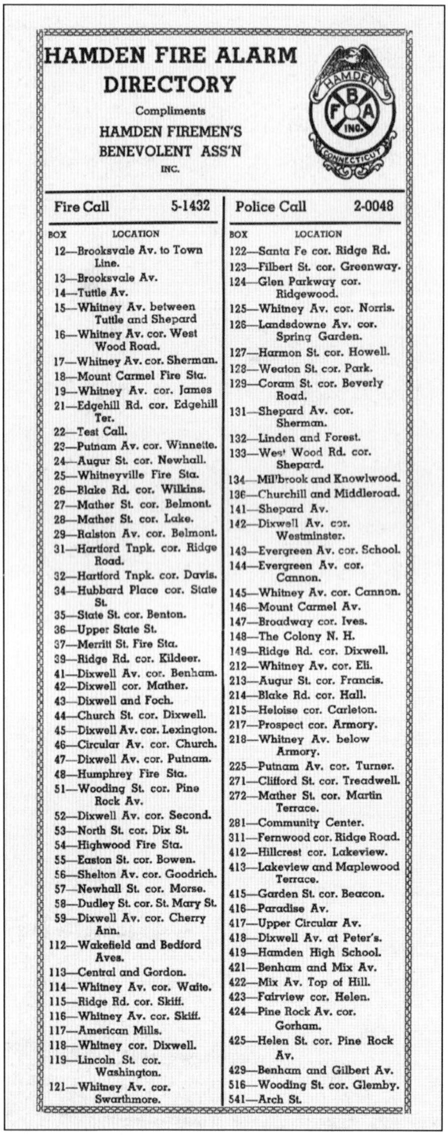

HAMDEN FIRE ALARM DIRECTORY

Compliments

HAMDEN FIREMEN'S BENEVOLENT ASS'N INC.

HAMDEN F B A INC. CONNECTICUT

Fire Call 5-1432

BOX — LOCATION

12—Brooksvale Av. to Town Line.
13—Brooksvale Av.
14—Tuttle Av.
15—Whitney Av. between Tuttle and Shepard
16—Whitney Av. cor. West Wood Road.
17—Whitney Av. cor. Sherman.
18—Mount Carmel Fire Sta.
19—Whitney Av. cor. James
21—Edgehill Rd. cor. Edgehill Ter.
22—Test Call.
23—Putnam Av. cor. Winnette.
24—Augur St. cor. Newhall.
25—Whitneyville Fire Sta.
26—Blake Rd. cor. Wilkins.
27—Mather St. cor. Belmont.
28—Mather St. cor. Lake.
29—Ralston Av. cor. Belmont.
31—Hartford Tnpk. cor. Ridge Road.
32—Hartford Tnpk. cor. Davis.
34—Hubbard Place cor. State St.
35—State St. cor. Benton.
36—Upper State St.
37—Merritt St. Fire Sta.
39—Ridge Rd. cor. Kildeer.
41—Dixwell Av. cor. Benham.
42—Dixwell cor. Mather.
43—Dixwell and Foch.
44—Church St. cor. Dixwell.
45—Dixwell Av. cor. Lexington.
46—Circular Av. cor. Church.
47—Dixwell Av. cor. Putnam.
48—Humphrey Fire Sta.
51—Wooding St. cor. Pine Rock Av.
52—Dixwell Av. cor. Second.
53—North St. cor. Dix St.
54—Highwood Fire Sta.
55—Easton St. cor. Bowen.
56—Shelton Av. cor. Goodrich.
57—Newhall St. cor. Morse.
58—Dudley St. cor. St. Mary St.
59—Dixwell Av. cor. Cherry Ann.
112—Wakefield and Bedford Aves.
113—Central and Gordon.
114—Whitney Av. cor. Waite.
115—Ridge Rd. cor. Skiff.
116—Whitney Av. cor. Skiff.
117—American Mills.
118—Whitney cor. Dixwell.
119—Lincoln St. cor. Washington.
121—Whitney Av. cor. Swarthmore.

Police Call 2-0048

BOX — LOCATION

122—Santa Fe cor. Ridge Rd.
123—Filbert St. cor. Greenway.
124—Glen Parkway cor. Ridgewood.
125—Whitney Av. cor. Norris.
126—Landsdowne Av. cor. Spring Garden.
127—Harmon St. cor. Howell.
128—Weaton St. cor. Park.
129—Coram St. cor. Beverly Road.
131—Shepard Av. cor. Sherman.
132—Linden and Forest.
133—West Wood Rd. cor. Shepard.
134—Millbrook and Knowlwood.
136—Churchill and Middleroad.
141—Shepard Av.
142—Dixwell Av. cor. Westminster.
143—Evergreen Av. cor. School.
144—Evergreen Av. cor. Cannon.
145—Whitney Av. cor. Cannon.
146—Mount Carmel Av.
147—Broadway cor. Ives.
148—The Colony N. H.
149—Ridge Rd. cor. Dixwell.
212—Whitney Av. cor. Eli.
213—Augur St. cor. Francis.
214—Blake Rd. cor. Hall.
215—Heloise cor. Carleton.
217—Prospect cor. Armory.
218—Whitney Av. below Armory.
225—Putnam Av. cor. Turner.
271—Clifford St. cor. Treadwell.
272—Mather St. cor. Martin Terrace.
281—Community Center.
311—Fernwood cor. Ridge Road.
412—Hillcrest cor. Lakeview.
413—Lakeview and Maplewood Terrace.
415—Garden St. cor. Beacon.
416—Paradise Av.
417—Upper Circular Av.
418—Dixwell Av. at Peter's.
419—Hamden High School.
421—Benham and Mix Av.
422—Mix Av. Top of Hill.
423—Fairview cor. Helen.
424—Pine Rock Av. cor. Gorham.
425—Helen St. cor. Pine Rock Av.
429—Benham and Gilbert Av.
516—Wooding St. cor. Glemby.
541—Arch St.

The Gamewell Alarm telegraph alarm system was once a vital fire service communications method, especially in the earliest days when residential telephone service was the exception and not the rule. In a Class B system, such as Hamden's before 1984, residents could summon the fire department by activating a fire alarm box (below) that would transmit a code to all stations that corresponded to the box location. The apparatus assigned to that box would respond. Before two-way radios, incident commanders could tap in a code from the closest alarm box to order additional alarms. Hamden Firemen's Benevolent Association, organized in 1932, distributed these lists of Hamden's fire alarm boxes to residents and businesses in the late 1930s. Hamden went to a Class A alarm system in 1984, which put responsibility for dispatching fire alarm boxes solely with the fire dispatcher. With the increased use of cell phones, Hamden's Gamewell alarm system, always plagued by malicious false alarms, was removed in 1998. (Both, HFRA.)

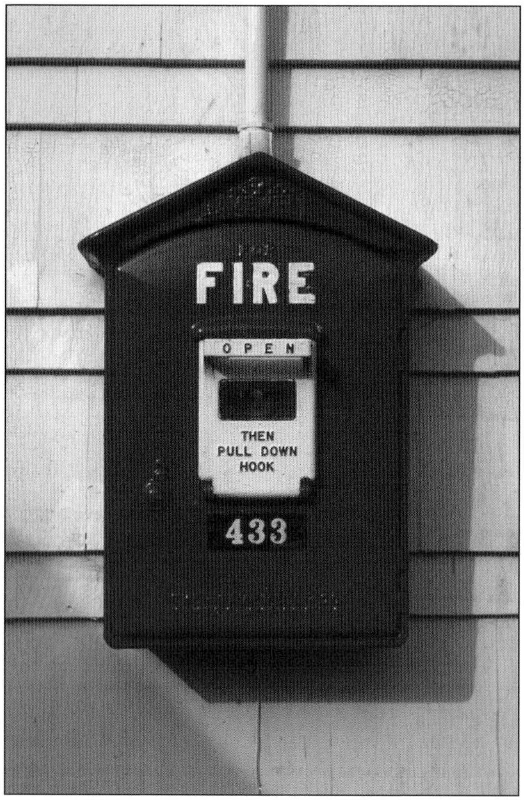

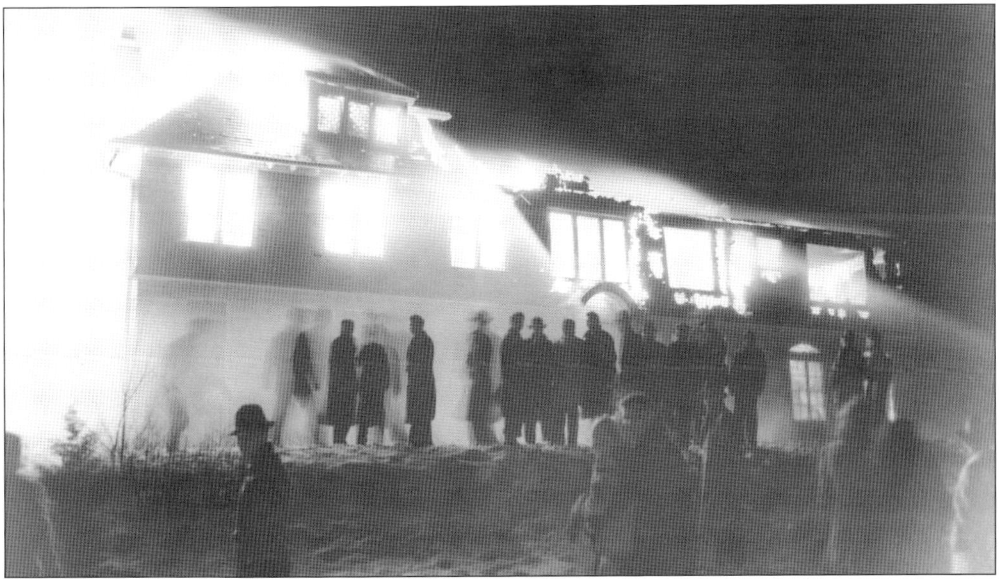

The old Peck School on Hillfield Road, vacant for several years, burned to the ground in the early morning of January 11, 1941. This 13-room, two-story structure had been operated by the Peck sisters. After it closed, Alice Peck became the teacher at the one-room schoolhouse farther up the street at the corner of Johnson and Still Hill Roads. (*New Haven Journal-Courier.*)

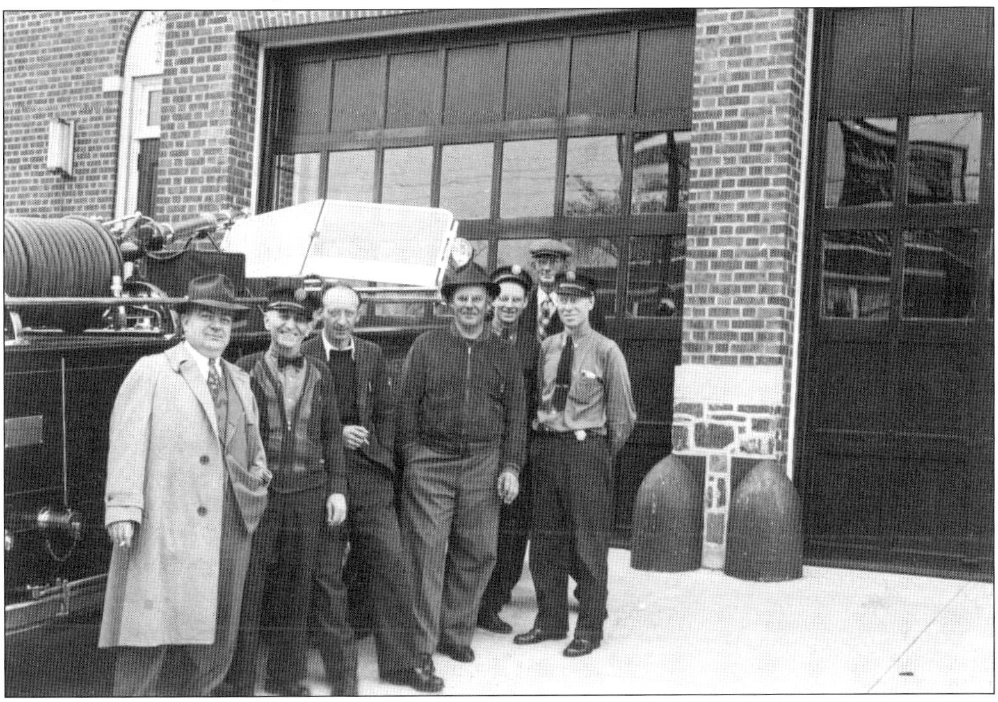

The gentlemen in this photograph taken at Centerville Station 4 in 1942 are, from left to right, Assistant Chief August Ball, Firefighter Joseph Marchitto (also known as "Joe Marks"), two unidentified men, Firefighter Clement Wetmore, unidentified, and Firefighter Roland Ruwet. Shortly after this photograph was taken, Ruwet became one of the department's first paid company officers. Note the recently installed rear step windshield on the 1938 Diamond-T Squad. (HFRA.)

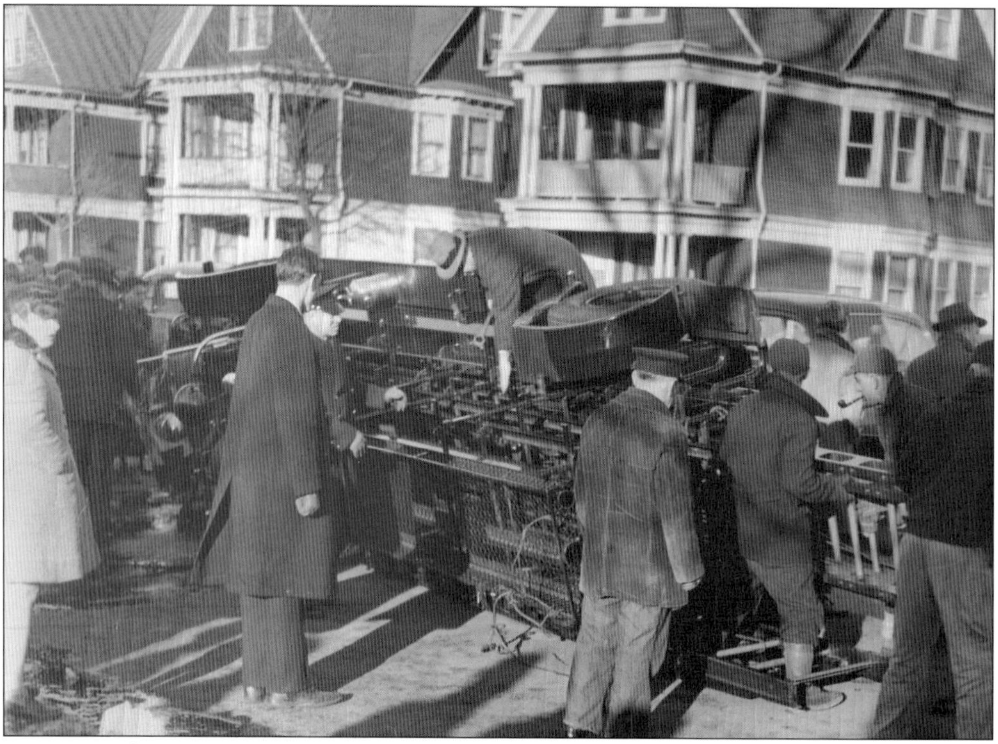

Driver David F. Howe was responding to a fire at the Peters' residence, where the Hamden Mart is today, when this 1926 Maxim ladder truck he was driving was struck by a trolley car at Dixwell and Mather on March 31, 1941. Firefighter Howe was uninjured, but the truck was deemed unsalvageable. Hamden ordered a new ladder truck from Wood Engineering Services that was delivered in early 1942. (I.A Sneiderman.)

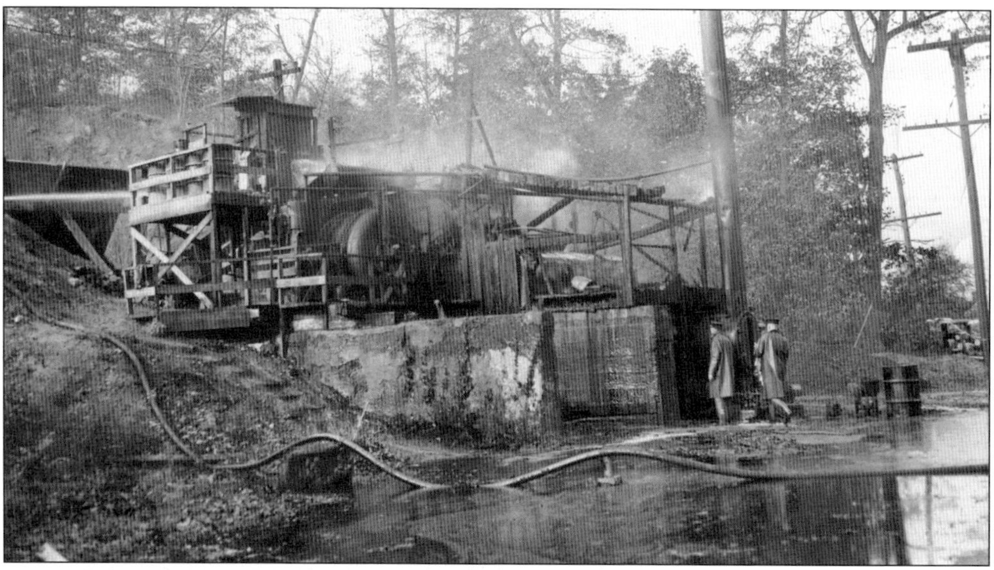

Hot coals ignited this fire at C.W. Blakeslee's Pine Rock Avenue mixing plant on November 3, 1941, which resulted in $10,000 in damages to the building and facial injuries to one of the employees. (I.A. Sneiderman.)

Three

The "Paid" Department

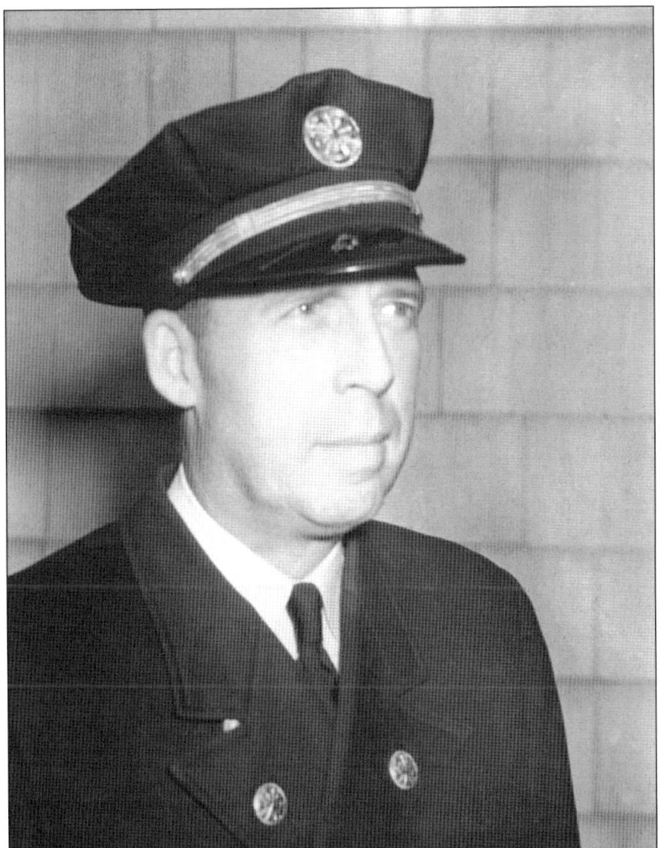

Following Pearl Harbor, the town needed full-time paid fire department officers. In April 1942, examinations were conducted for fire chief and captain. Having scored highest on the chief's exam, Raymond C. Spencer, seen here, was selected as Hamden's first paid fire chief. Scoring highest on the captain's exam, Joseph Hromadka and Albert Purce were appointed captains. Runners-up Everett Doherty and Roland Ruwet were appointed lieutenants. (Fire Chief's Office.)

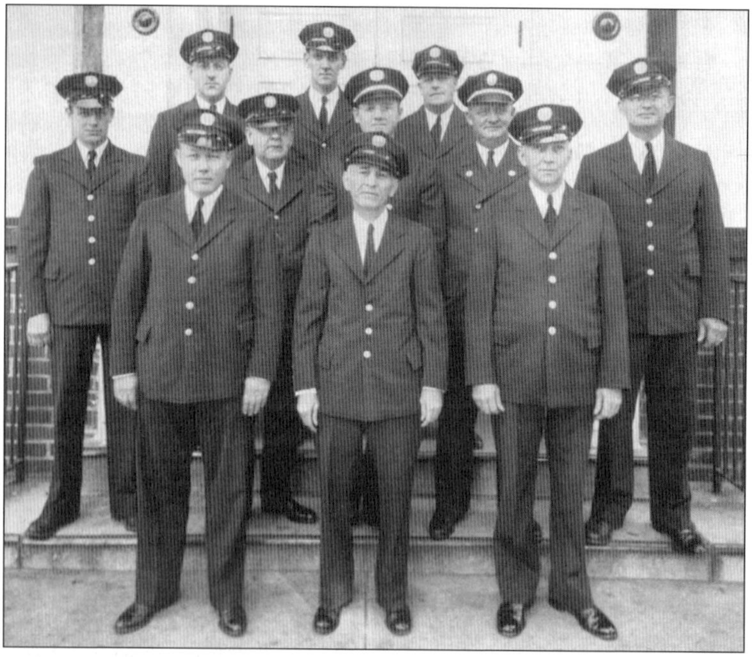

Following the 1942 department reorganization, the personnel of Platoon 1 posed at Whitneyville Station 3. From left to right are (first row) Daniel Hume, Joseph Marchitto, and Frank Nolan; (second row) Clement Wetmore, Albert Molleur, Capt. Joseph Hromadka, Lt. Everett Doherty, and George Thatcher; (third row) Stewart Keeler, James Strain, and Arthur Norman. (HFRA.)

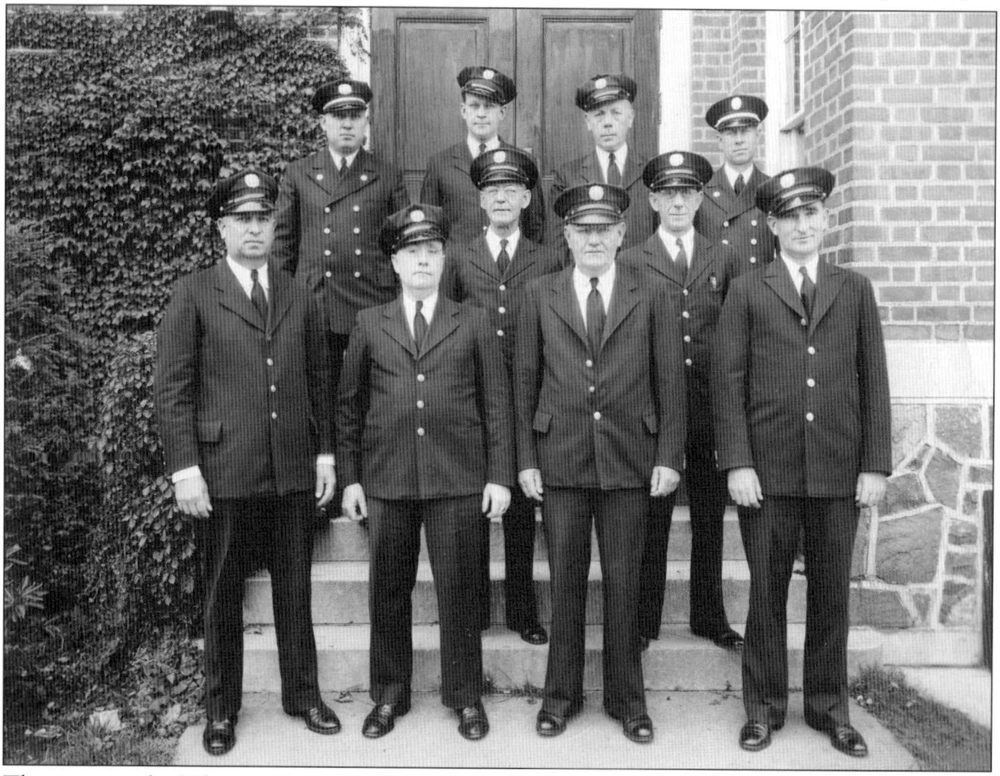

The personnel of Platoon 2 pose at the town hall steps. From left to right are (first row) Julius Norman, Clement Kammerer, Joseph Dukat, and Walter Thomas; (second row) David F. Howe and Ralph Rosson; (third row) Capt. Albert Purce, Emil Strain, Robert Reutenauer, and Lt. Roland Ruwet. Missing is V. Paul Leddy, who was serving in the US Army. (HFRA.)

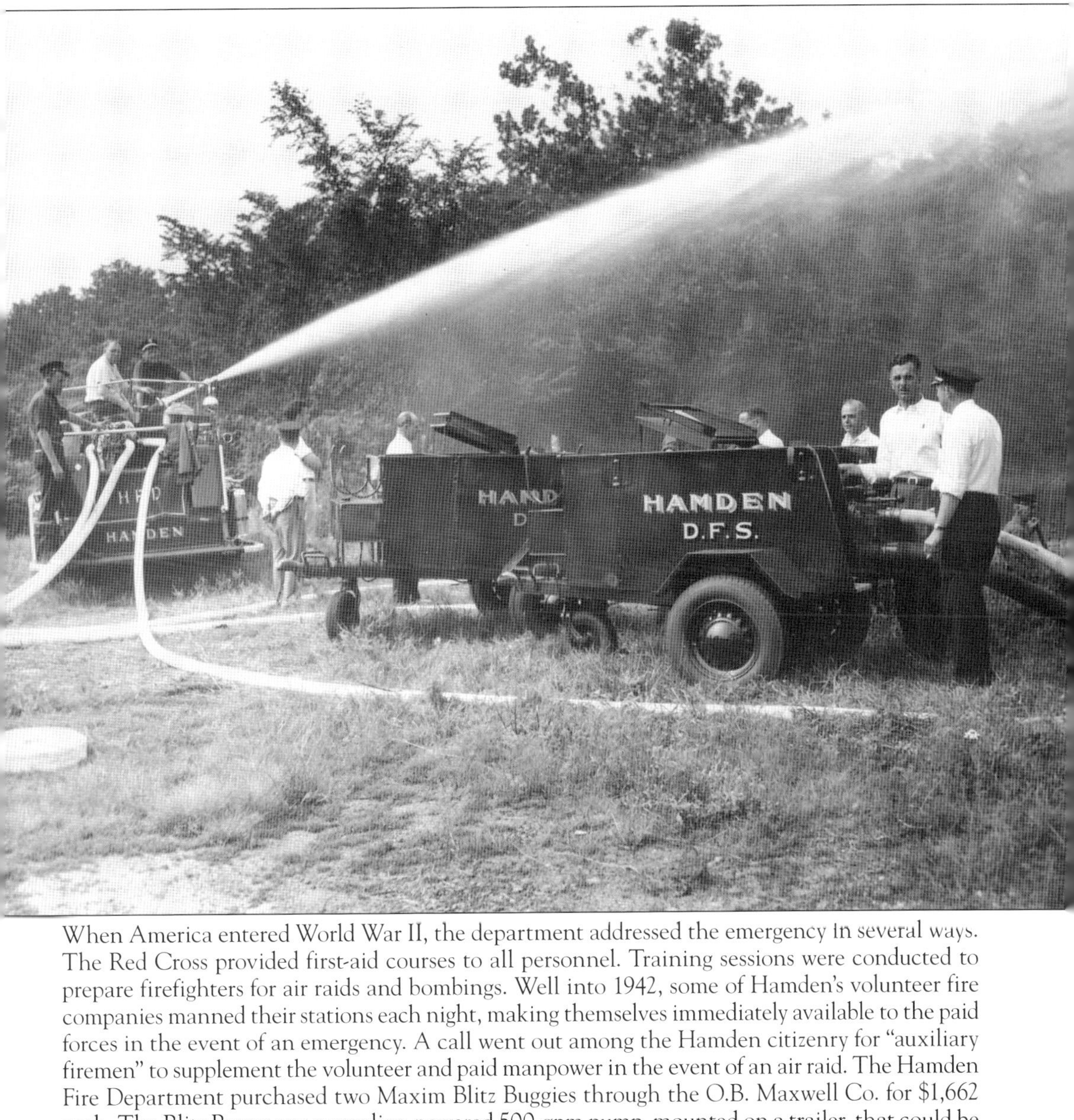

When America entered World War II, the department addressed the emergency in several ways. The Red Cross provided first-aid courses to all personnel. Training sessions were conducted to prepare firefighters for air raids and bombings. Well into 1942, some of Hamden's volunteer fire companies manned their stations each night, making themselves immediately available to the paid forces in the event of an emergency. A call went out among the Hamden citizenry for "auxiliary firemen" to supplement the volunteer and paid manpower in the event of an air raid. The Hamden Fire Department purchased two Maxim Blitz Buggies through the O.B. Maxwell Co. for $1,662 each. The Blitz Buggy was a gasoline-powered 500-gpm pump, mounted on a trailer, that could be hand-drawn or towed to a fire scene. It discharged up to 500 gallons of water per minute, either from a hydrant or by drafting from a static water source. One of the Blitz Buggies was stationed in a garage in Spring Glen, the other in the Wilmot section of Pine Rock. (I.A. Sneiderman.)

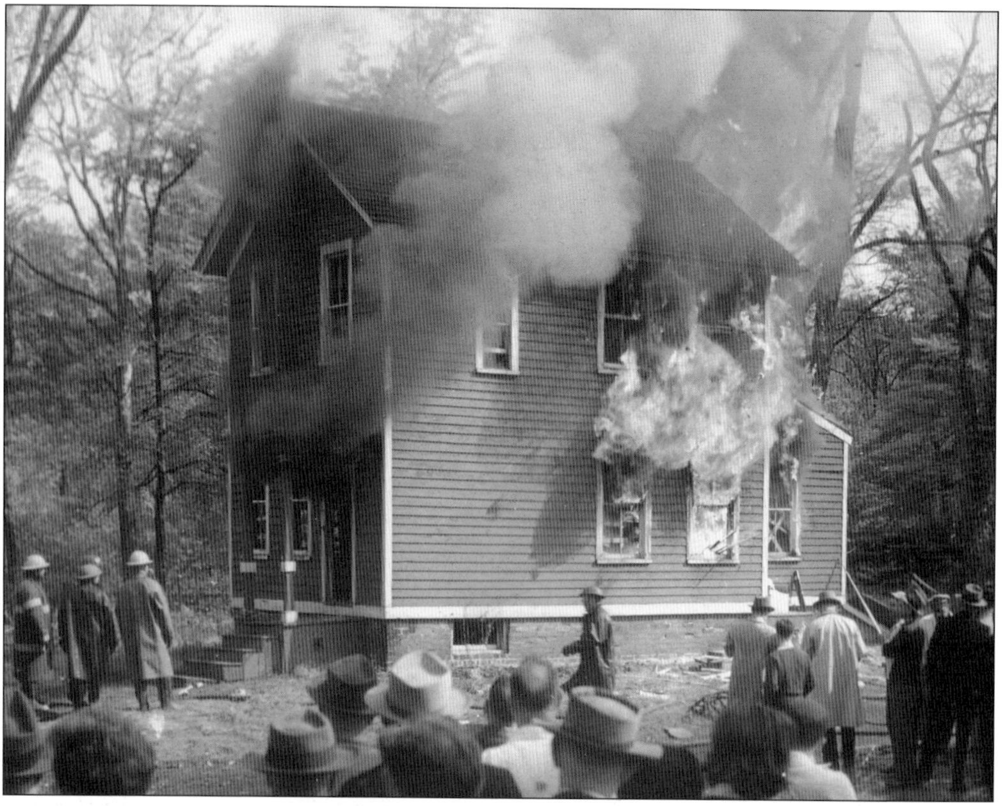

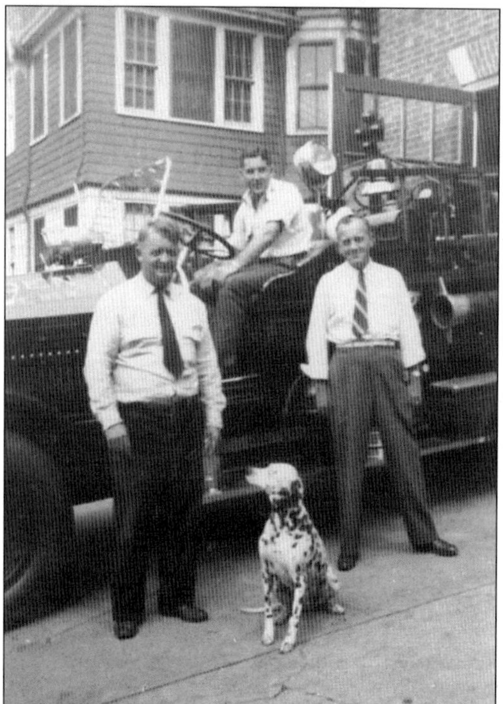

The department set fire to this derelict house behind the old Woodruff factory on Ives Street to test a new fog nozzle manufactured by the Rockwood Company. The October 1945 test, a joint exercise by the fire departments of Hamden and New Haven, was written up in *Fire Engineering* magazine. (I.A. Sneiderman.)

Firefighter Robert "Bubby" O'Donnell, behind the wheel of Co. 3's 1928 Maxim, is the department's newest career member in this 1947 photograph taken at Whitneyville Station 3. Whitneyville's station mascot, a beautiful Dalmatian named Belle, poses with Lt. Everett Doherty (left) and the Whitneyville Volunteer Co. 3 captain, Lawrence Spahr. Belle rode to all calls in her own special seat on Engine 3. (Thomas Doherty.)

In this August 14, 1947, photograph, Chief Raymond Spencer demonstrates the department's new two-way "car link" radio. Few departments were completely radio linked in those days, and it would be several more years before every vehicle in the department was equipped with two-way radios. (I.A. Sneiderman.)

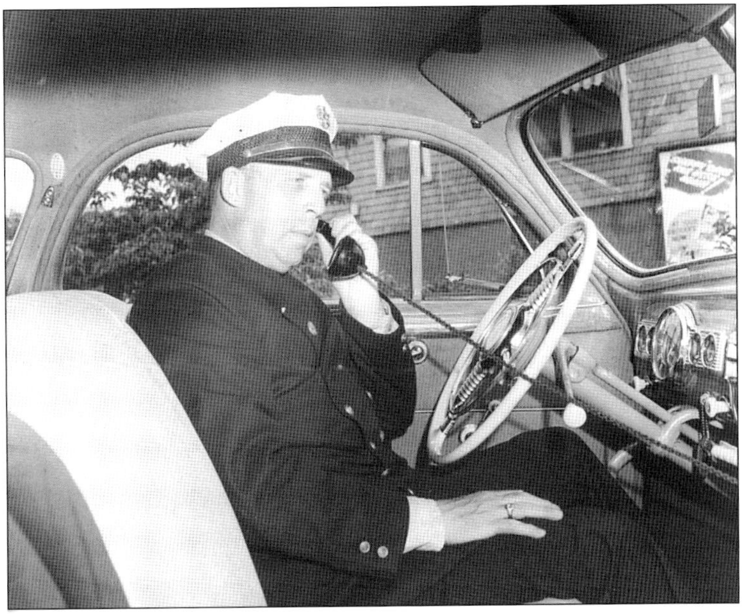

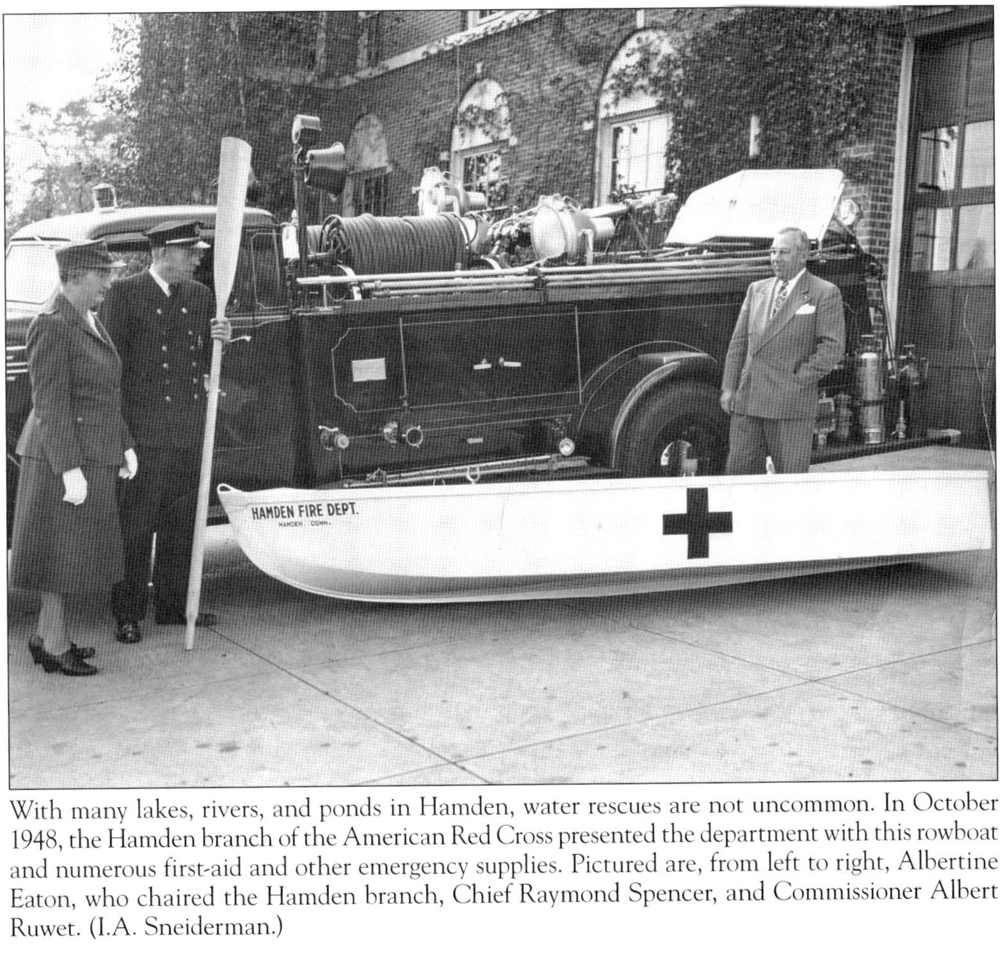

With many lakes, rivers, and ponds in Hamden, water rescues are not uncommon. In October 1948, the Hamden branch of the American Red Cross presented the department with this rowboat and numerous first-aid and other emergency supplies. Pictured are, from left to right, Albertine Eaton, who chaired the Hamden branch, Chief Raymond Spencer, and Commissioner Albert Ruwet. (I.A. Sneiderman.)

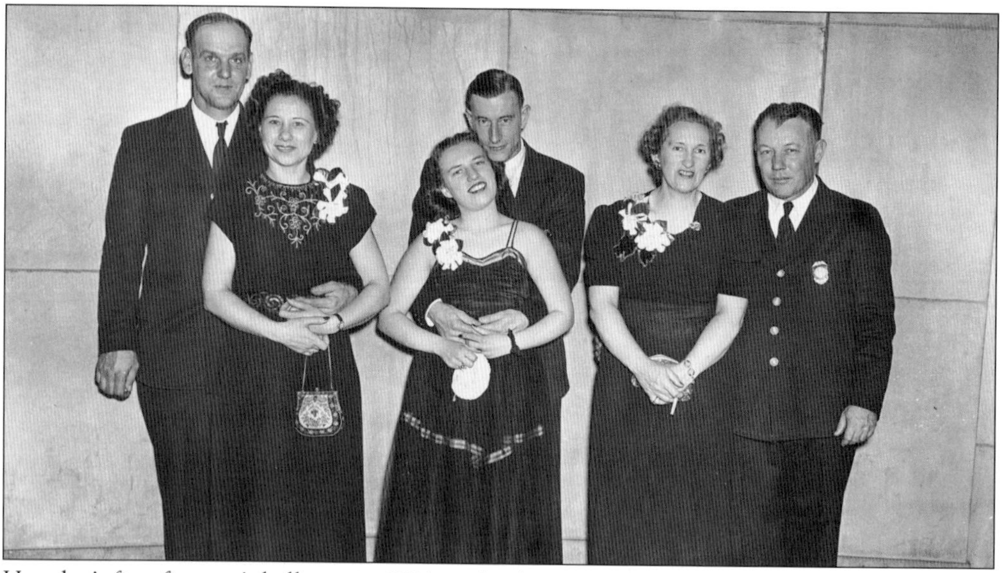

Hamden's first fireman's ball was in 1934, and it became an annual tradition that lasted until 1974. At the 1948 Hamden Firemen's Ball, the three firefighters and their wives pictured here are, from left to right, James and Lillian Strain, Ernestine and William Hines, and Kathryne and Daniel Hume. (Sally Hines-Row.)

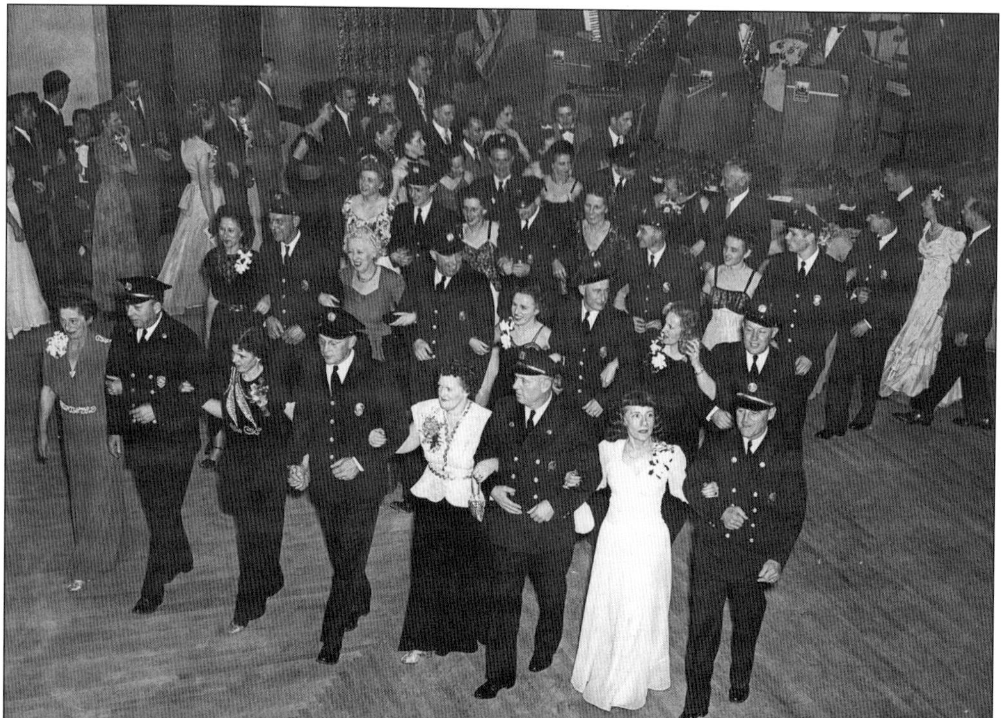

The Grand March kicked off every annual ball. It was customary in those days for members to wear their dress uniforms. Escorting their ladies at the front of the march are, from left to right, Chief Raymond C. Spencer, Firefighter Emil Strain, Capt. Albert Purce, and Capt. Joseph Hromadka. (Sally Hines-Row.)

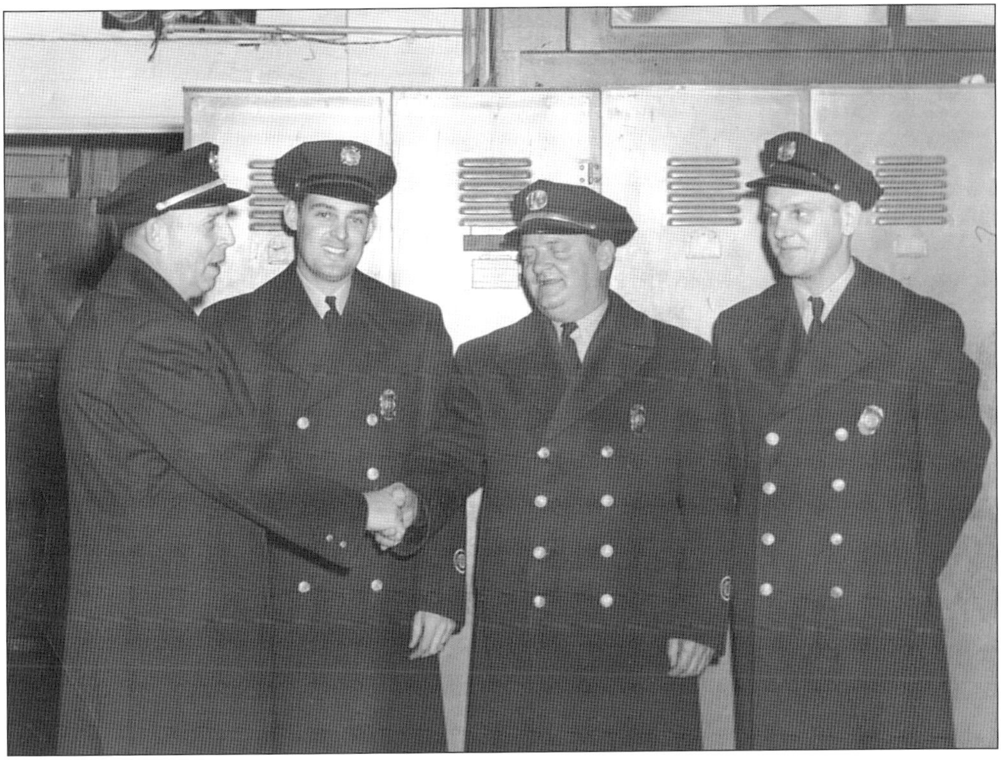

When Capt. Albert Purce was appointed the town's first full-time fire marshal in 1949, Everett Doherty was promoted to captain. In this December 1949 photograph, Captain Doherty is congratulated by Chief Spencer (left), along with newly appointed lieutenants Paul Leddy (second from left) and James Strain. (Sgt. George Moehl, Hamden Police Department.)

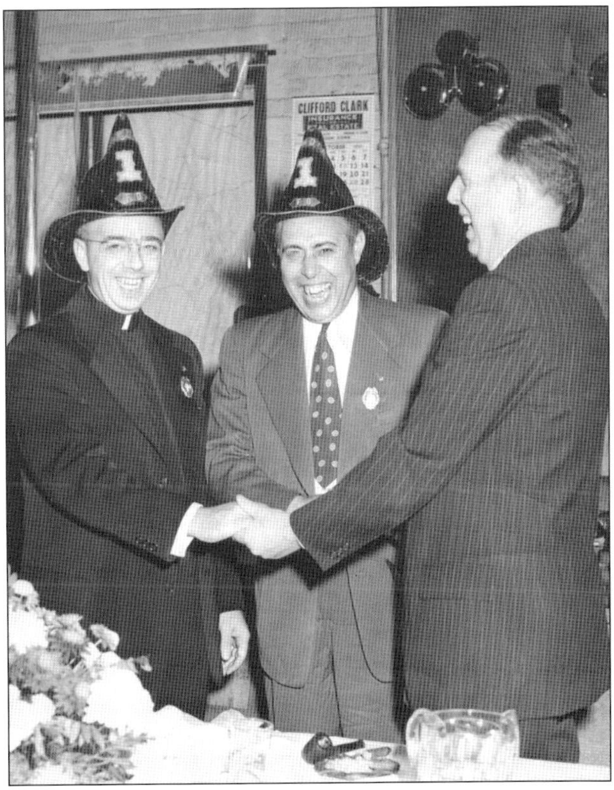

The department's first chaplains, Fr. Bernard Miller (left) of St. Rita's Roman Catholic Church in Spring Glen and Rev. Joseph Peters of the Mount Carmel Congregational Church, are congratulated by Chief Spencer. The two chaplains were installed at the annual meeting of the Hamden Paid Firemen's Sick Benefit Association on October 10, 1950. (Paul Wetmore Sr.)

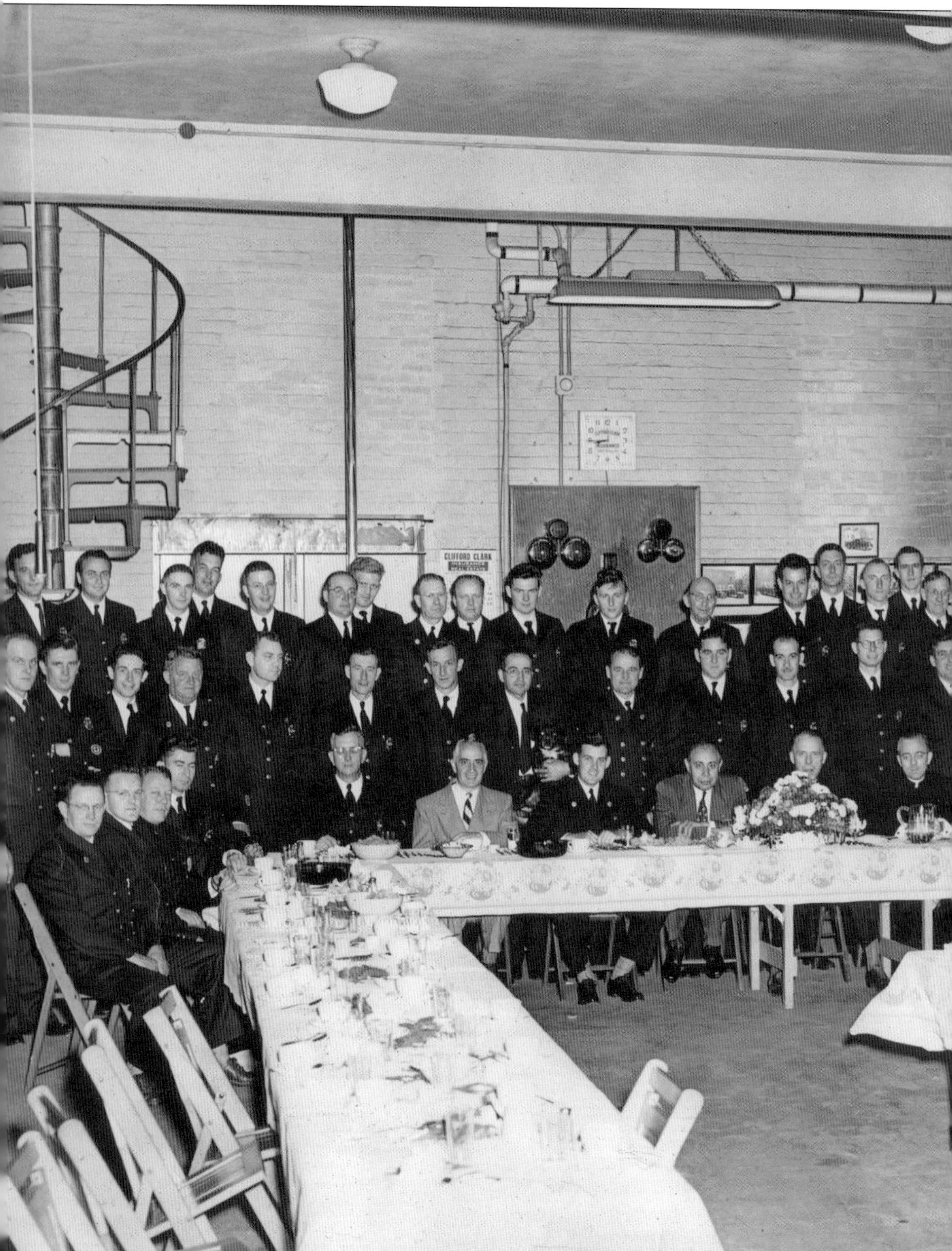
CLIFFORD CLARK

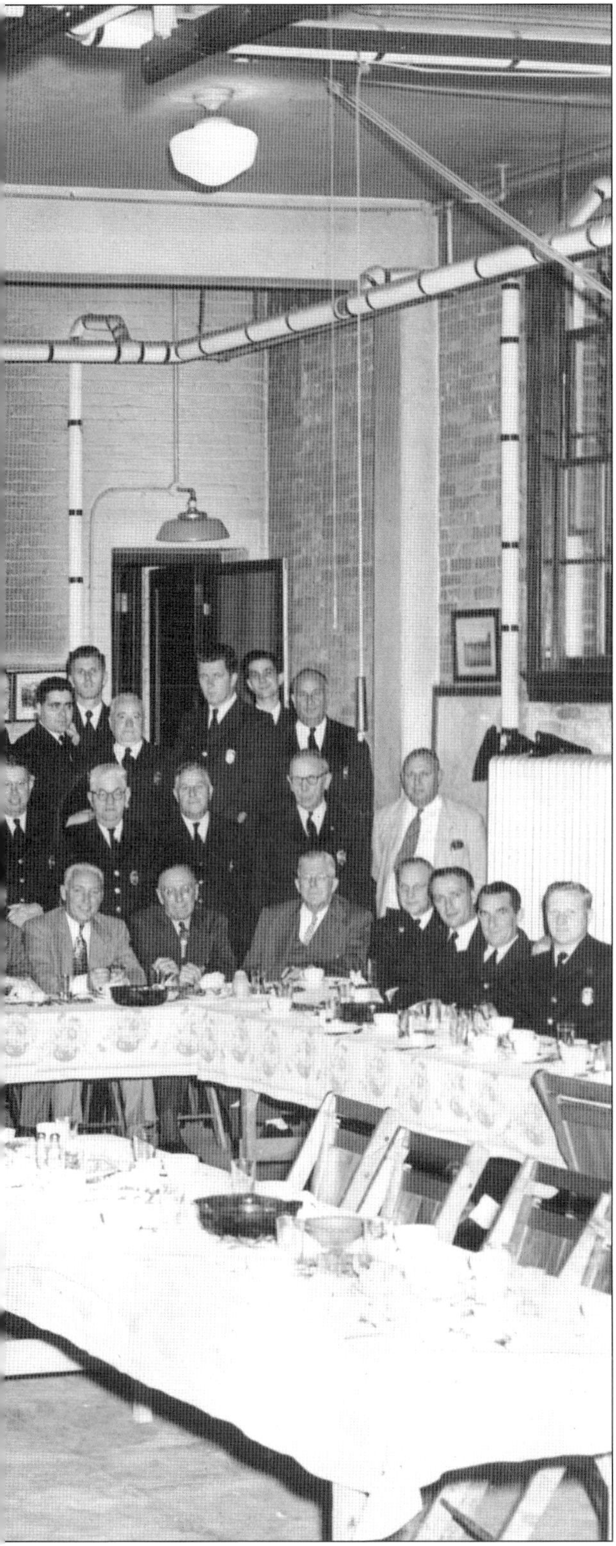

All department personnel, except one, assembled at headquarters on October 10, 1950. From left to right are (first row, seated) Supt. Clement Wetmore, Howard Hurlburt, Lt. Daniel Hume, Paul Rosadina, Wilbur Baker, Third Selectman Albert Connolly, Lt. Paul Leddy, Rev. Joseph Peters (Chaplain), Chief Raymond C. Spencer, Fr. Bernard Miller (chaplain) Commissioner Oscar Lampe, Commissioner Herbert Hume, Commissioner Hudson Weigold, commission secretary Benjamin C. Bamford, Lt. Emil Strain, Raymond Carofano, Walter Thomas, and Luke Tobin; (second row) Lt. James Strain, Raymond Vreeland, Francis Leddy, Capt. Everett Doherty, Burton Hillocks, William Hines, Francis Eitler, Mario Serafino, Capt. Joseph Hromadka, Robert O'Donnell, William Bossoli, George Reutenauer, Vincent Roth, John Hoffman, Joseph Dukat, Arthur Norman, Robert Reutenauer, and Marshal Albert Purce; (third row) Edward Bevins, Russell Norman, John McKee, Arthur Smith, Lawrence Bellemore, Sidney Trower, Robert Williams, Lt. Roland Ruwet, David Herrmann, David Howe, John Laffin, Francis Nolan, Alfred Ramelli, Kenneth Harrington, Warren Blake, Thomas Cummins, Ralph Rosson, Frederick Fletcher, Salvatore Ferraro, Hugh McLean, Clement Kammerer, Daniel O'Connell, Richard Lostritto, and Julius Norman. Firefighter William Esposito was covering the south end of town at Station 2. (Lewis Berlepsch.)

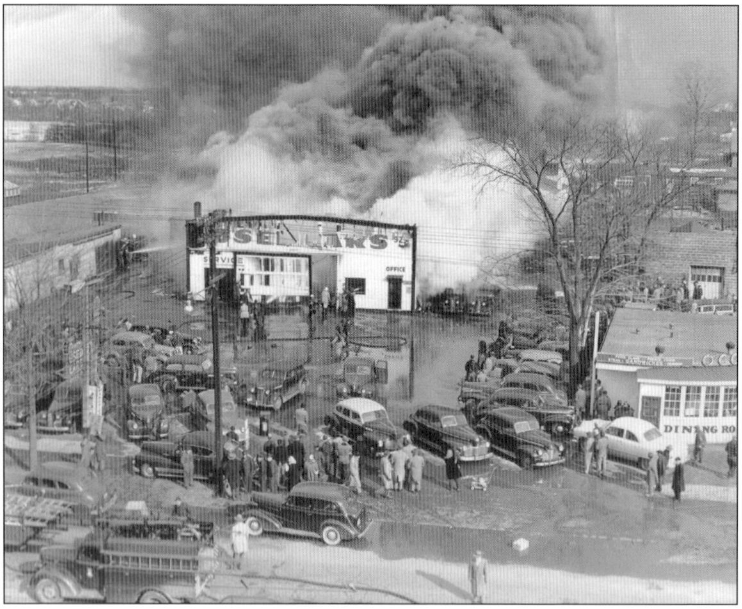

This photograph of the March 2, 1950, fire at Lynn Motors, 925 Dixwell Avenue, was taken by noted news photographer I.A. Sneiderman from the bell tower of St. Ann's Church at the corner of Dixwell Avenue and Arch Street. This nasty blaze burned several buildings and consumed dozens of old tires as well as several used cars that any collector would cherish today. (I.A. Sneiderman.)

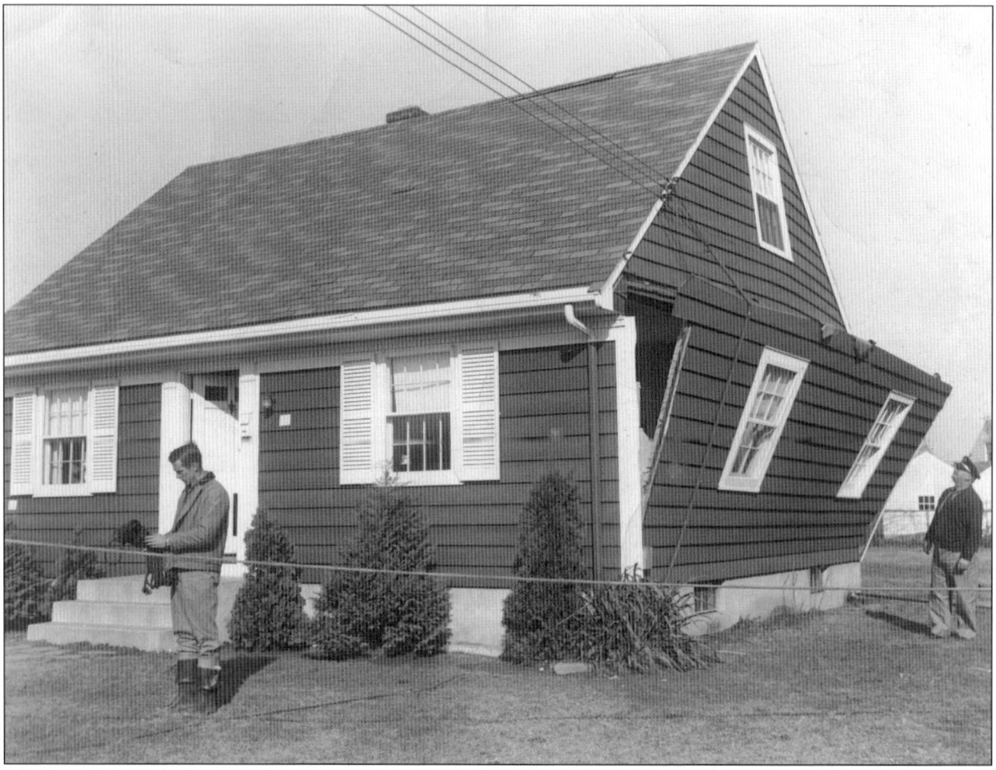

A November 1950 gas explosion that took the life of a young mother blew out the walls of this new home on Thayer Street in the Pine Rock area. Firefighter Thomas "Tim" Cummins (left) and Capt. Everett Doherty are pictured. (Thomas Doherty.)

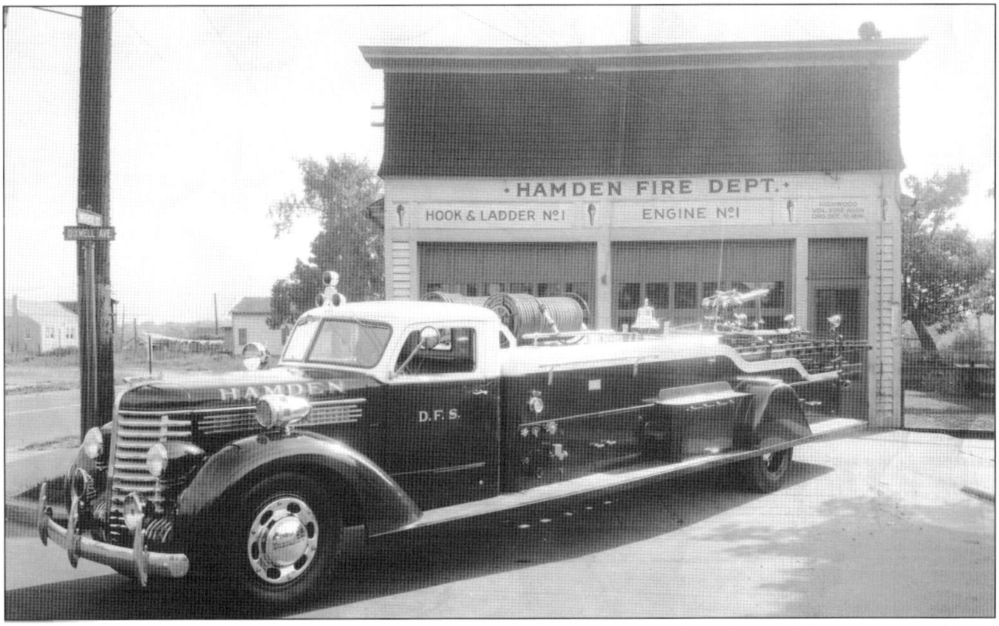

This 1941 Diamond-T city service ladder truck, parked in front of Highwood Station 1, replaced the 1926 Maxim ladder truck that was totaled in March 1941. When Station 1 closed on October 1, 1951, the Highwood volunteers disbanded, and the building was sold. This ladder truck and Engine 1, a 1942 Diamond-T 600-gpm pumper, were transferred to Humphrey Station 2, only six-tenths of a mile away. (HFRA.)

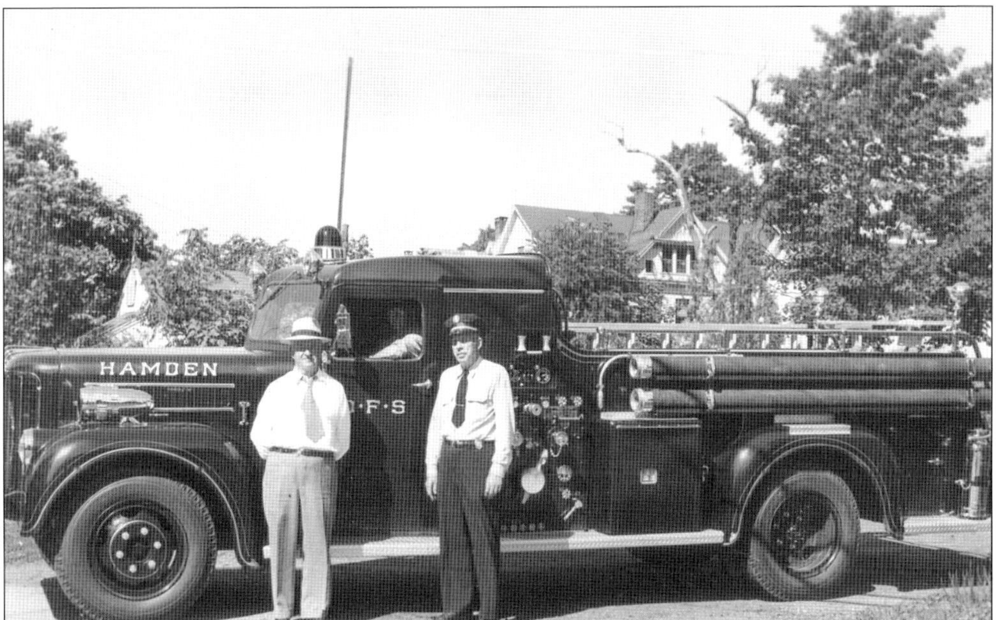

Delivered in August 1951, this Maxim 750-gpm pumper was the department's first new postwar apparatus. On the ramp at Humphrey Station 2 are, from left to right, Patrick Monahan of the O.B. Maxwell Company, Firefighter Wilbur Baker (behind the wheel), and Chief Raymond C. Spencer. (I.A. Sneiderman.)

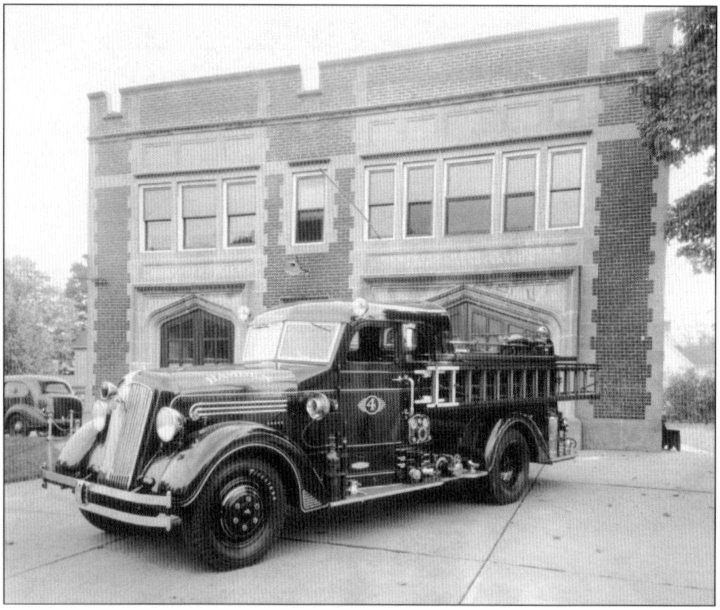

This 1942 photograph of the 1938 Seagrave at Humphrey Station 2 on Circular Avenue shows how the building was originally configured, with one full-size apparatus bay and a smaller bay for a hose cart. A 1950 National Board of Fire Underwriters report recommended closing Highwood Station 1 and moving its apparatus to Humphrey Station 2, which was renovated in 1951 to its present three-bay configuration. (HFRA.)

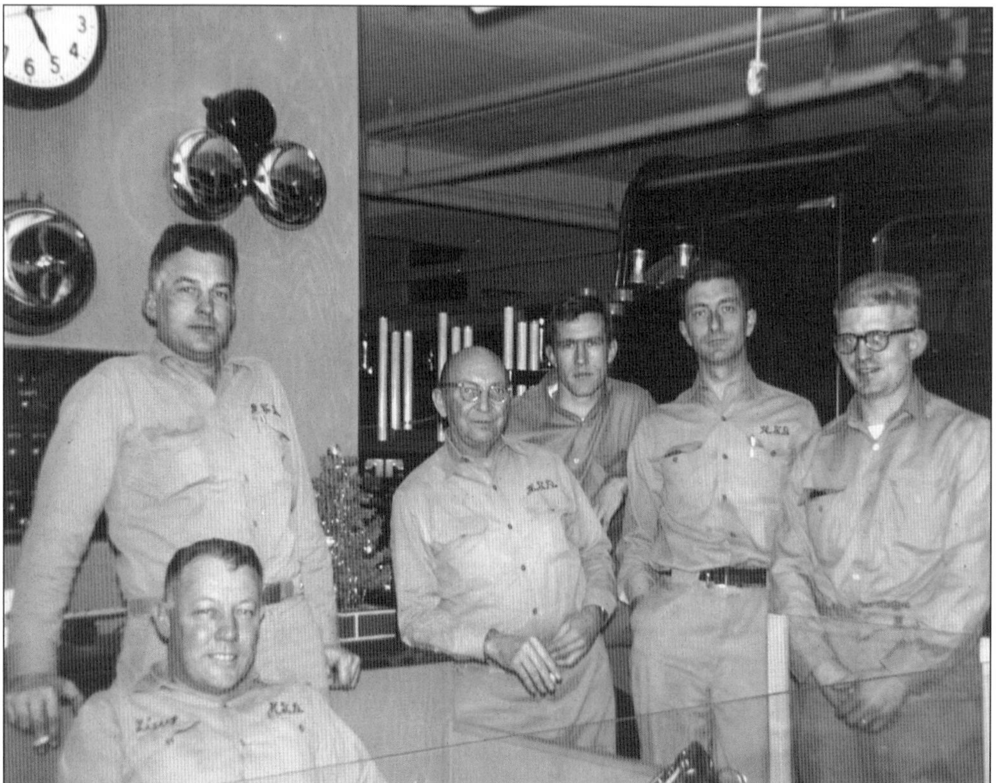

In December 1951, the Platoon 2 personnel at Station 2 are, from left to right, Lt. Daniel Hume (seated) and Firefighters Arthur Smith, Francis Nolan, Thomas Cummins, Kenneth Harrington, and Robert "Whitey" Williams. The department's new Engine 2, a Maxim 750 pumper, can be seen behind them in the middle bay. (HFRA.)

On a frigid morning in February 1952, Hamden's Squad and Engine 5 responded with five other area fire departments on a mutual aid call to the Waverly Inn on Maple Avenue in Cheshire for a fire that did $100,000 in damages. (I.A. Sneiderman.)

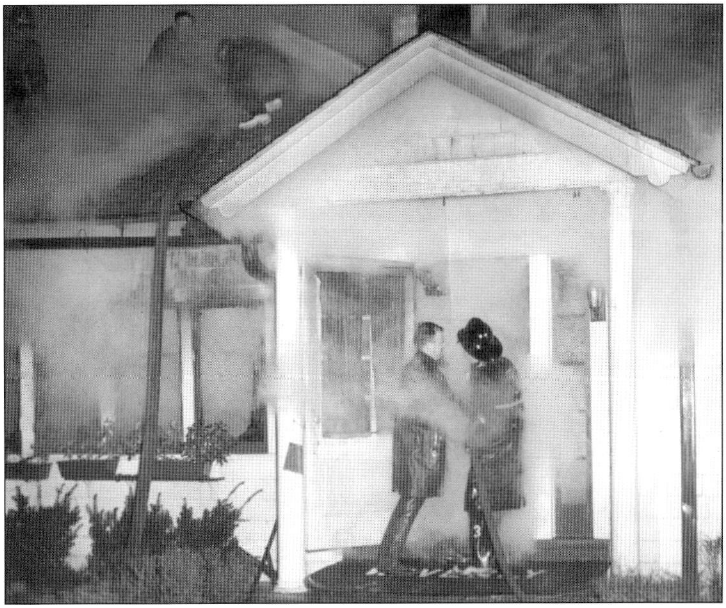

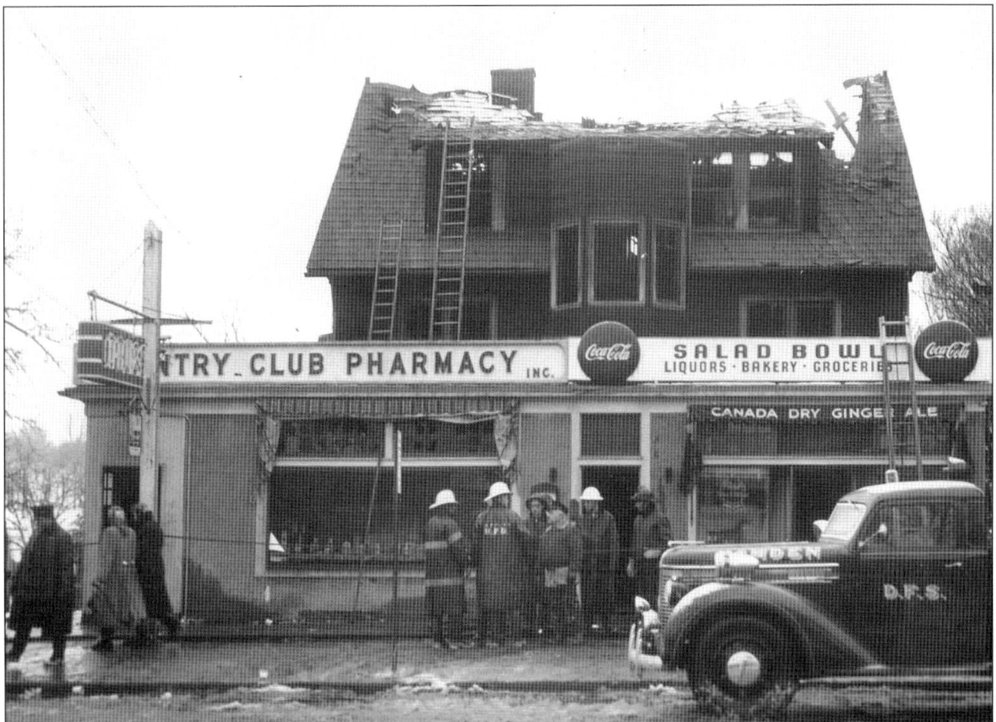

On February 12, 1953, an early morning fire extensively damaged the Country Club Pharmacy on the southeast corner of Whitney and Putnam Avenues. According to the *Second Annual Hamden Firemen's Memorial Service* program, dated June 10, 1934, this was the site of a barn where the Whitneyville fire company held its organizational meeting in 1910. (I.A. Sneiderman.)

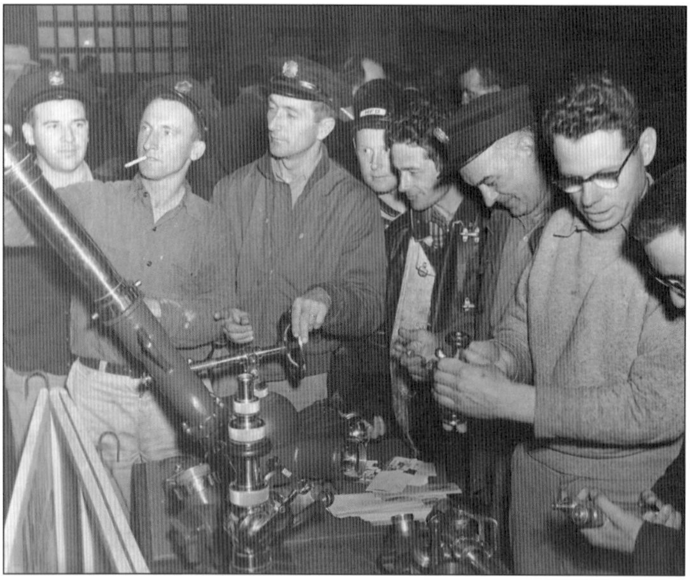

From left to right, Firefighters Joseph Collins, Warren Blake, Kenneth Harrington, John O'Hare, Francis Leddy, Arthur Smith, George Reutenauer, and William Esposito examine a new-style desk gun at the New Haven Fire School in May 1954. Harrington would become deputy chief/training officer 20 years later. Leddy and Reutenauer would also rise to the rank of deputy chief (shift commander). (I.A. Sneiderman.)

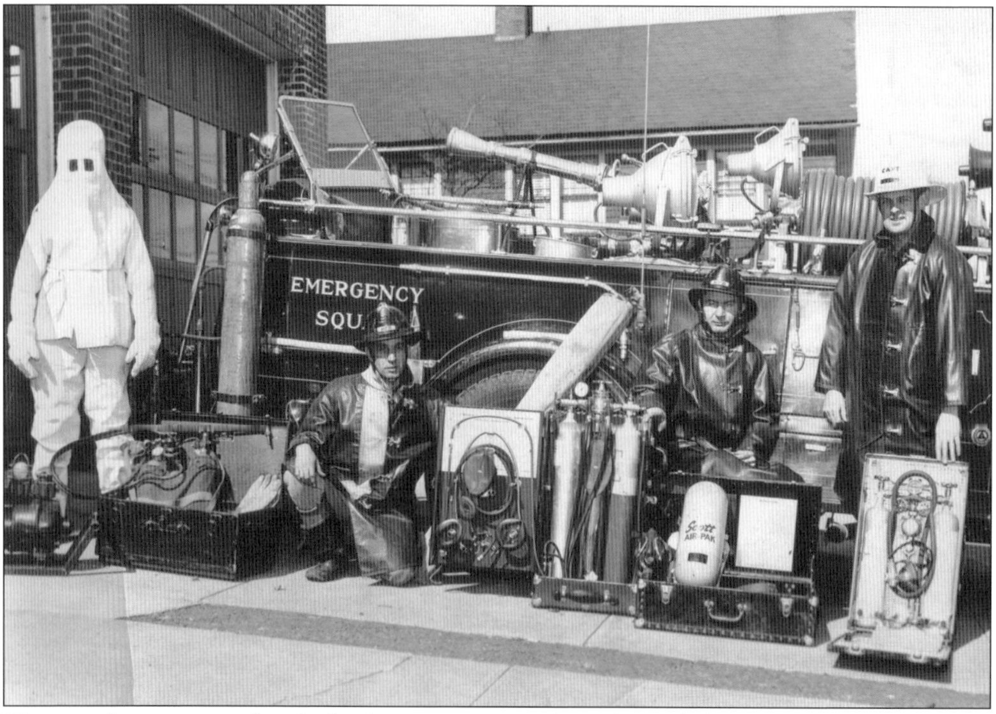

The squad was equipped with the latest equipment for 1954, including a 1,250-watt generator, two floodlights, portable acetylene cutting torch, stretcher, Scott Air Pack, H and H Inhalator, and an E and J Resuscitator. But the walking hazmat incident in this photograph is Firefighter William Bossoli (left) in the asbestos suit. Others are, from left to right, Firefighters Richard Lostritto and Paul Rosadina and Capt. Paul Leddy. (*New Haven Register.*)

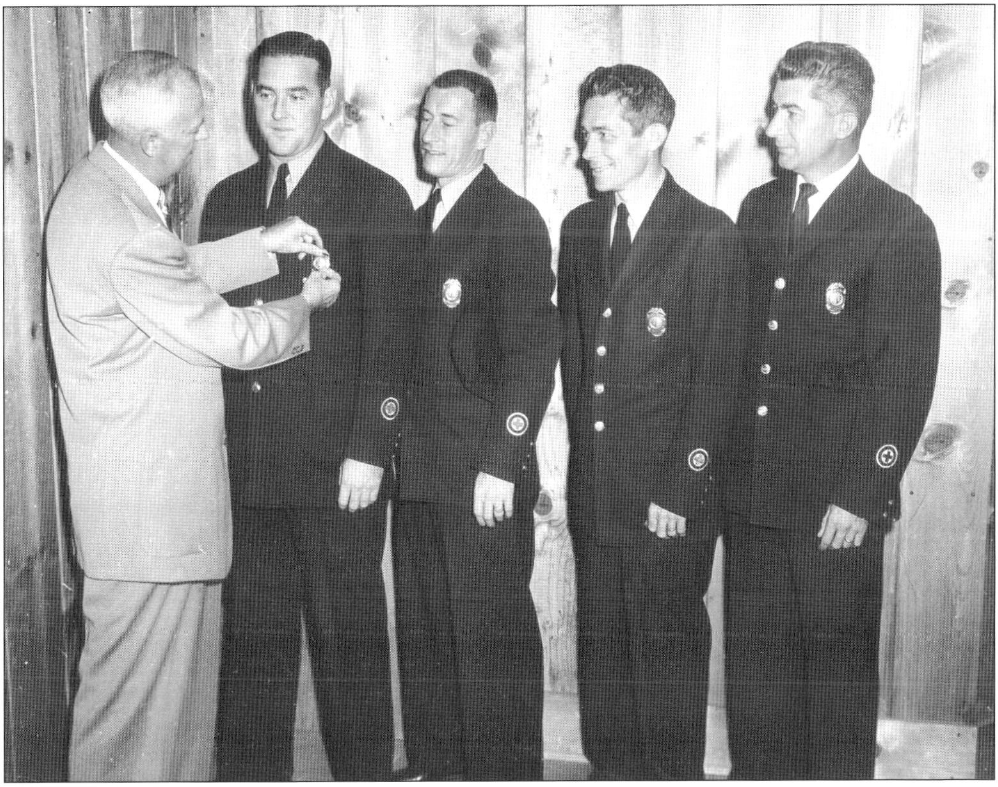

In 1956, four veteran firefighters were promoted to fill all department lieutenant slots, which had been vacant for two years. First Selectman Herbert Hume (left) pins lieutenant badges on, from left to right, Robert O'Donnell, William Hines, Francis Leddy, and Paul Rosadina. Lieutenant O'Donnell was promoted to captain six months later, when Capt. Daniel Hume was appointed the department's first deputy chief/training officer. (I.A. Sneiderman.)

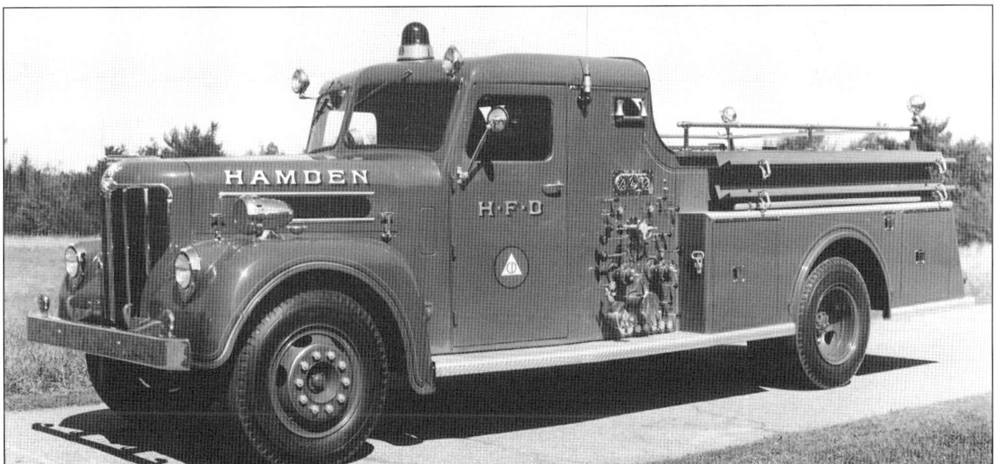

The Civil Defense logo on the doors of Hamden's new 1954 Maxim 750-gpm pumper signifies that half the cost of this $17,800 apparatus was paid with a grant from the federal government. The pumper was assigned to Centerville Station 4 and served the department in various capacities until well into the 1990s. (Chandler Brainard.)

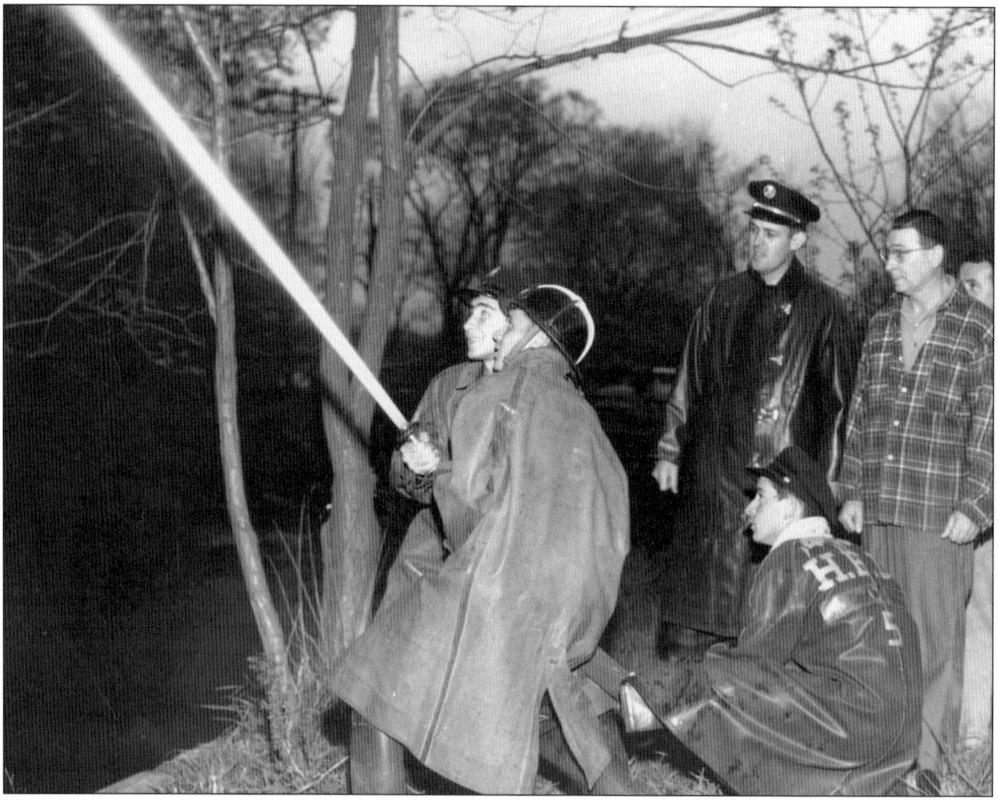

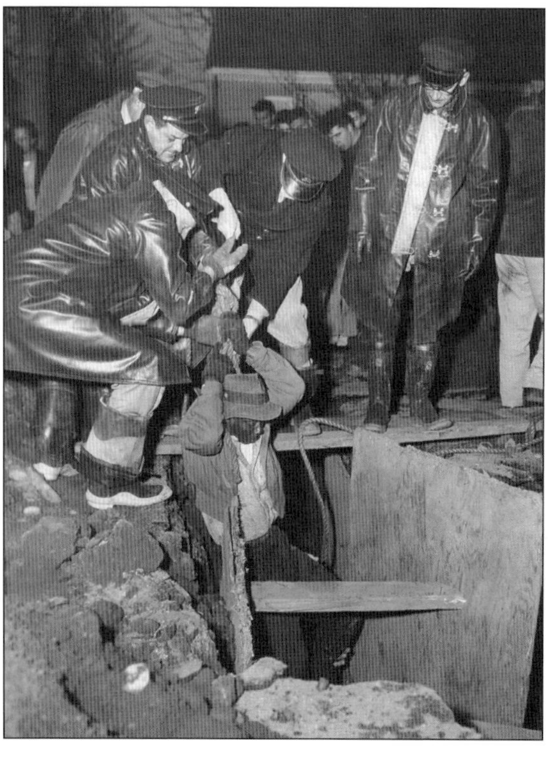

Organized in March 1955, Hamden's Civil Defense Fire Auxiliary was comprised mainly of Hamden volunteer firefighters. Battalion Chief V. Paul Leddy, in dark bunker coat, instructs the volunteers on handling a fully charged two-and-a-half-inch hand line. The volunteers are, from left to right, Benjamin Feinn, Merton "Bill" McAvoy Jr., John Tramontano, and Fred Clarke (standing). Twenty-nine years later, Tramontano would succeed Leddy as Hamden's fire chief. (I.A. Sneiderman.)

On December 1, 1955, the personnel of Platoon 3 conducted this below grade rescue of a worker trapped in a trench on Hayward Road. From left to right are Firefighters Julius Norman and Alfred Ramelli, Battalion Chief V. Paul Leddy, and Firefighter William Bossoli. (I.A. Sneiderman.)

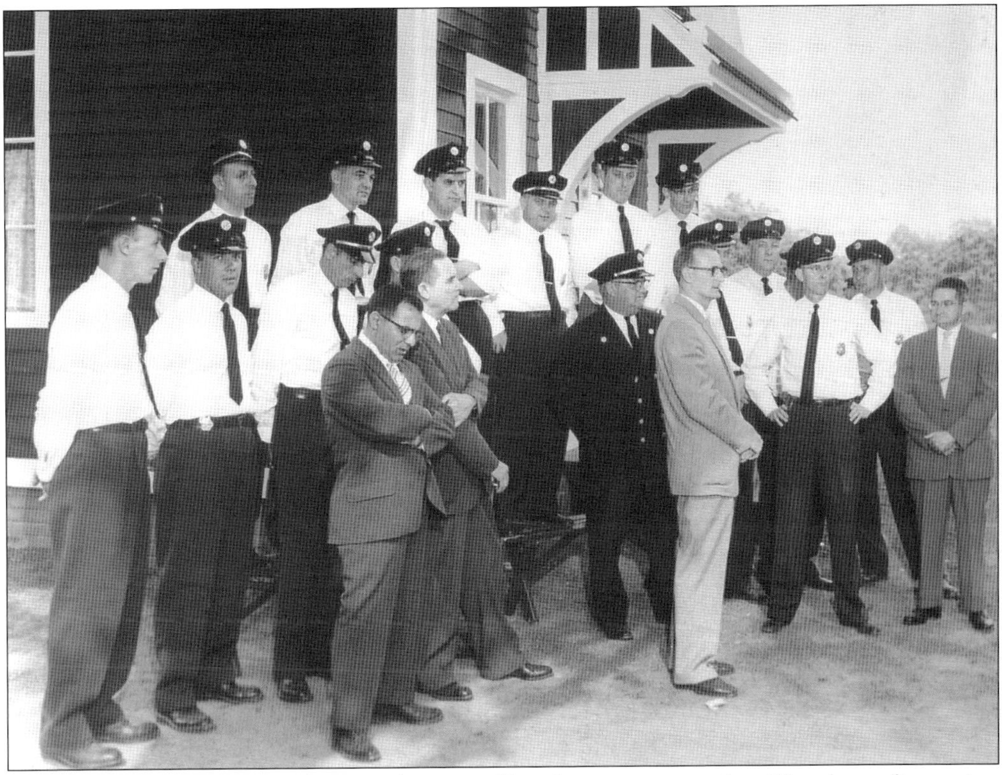

By the mid-1950s, West Woods in northwestern Hamden was growing fast. The closest fire station was several miles away in Mount Carmel. In 1956, Battalion Chief Everett Doherty (in dress blues) helped organize the West Woods Volunteer Fire Co. No. 9. The former one-room schoolhouse at Johnson and Still Hill Roads was turned over to the new association in a ceremony pictured in this photograph. (I.A. Sneiderman.)

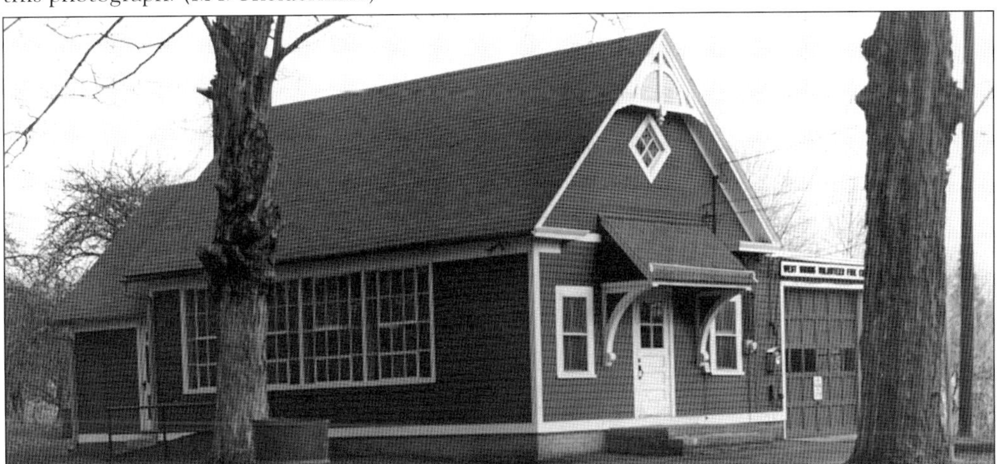

In 1958, the West Woods volunteers added an apparatus bay to their new quarters. Co. 9's first pumper was the 1925 Seagrave Suburbanite that started out as Engine 4 more than 30 years earlier. This structure was moved farther back on the property in 1967 to permit construction of the modern fire station that stands there today. The volunteer fire company at West Woods remained active until 1988. (I.A. Sneiderman.)

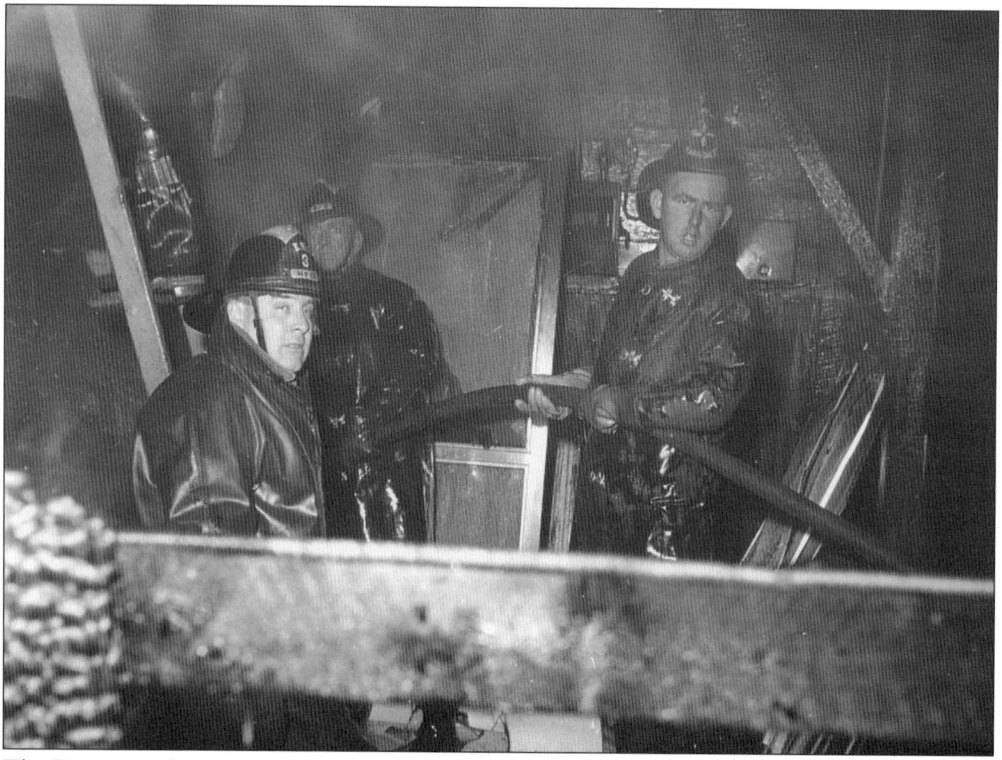

The Eastern Atlantic Lumber Company at Gorham and Circular Avenues was destroyed in a general alarm fire Monday night, November 11, 1957. Pictured in the photograph above are, from left to right, Firefighters Sidney Trower and Wilbur Baker, along with Volunteer Firefighter William Hindinger from Dunbar Hill. The aftermath of the blaze, seen below, shows the devastation. Subsequent investigations by the fire marshal's office and state officials concluded that the cause of the fire was incendiary in nature. (Above, photograph by Lubee; below, photograph by I.A. Sneiderman.)

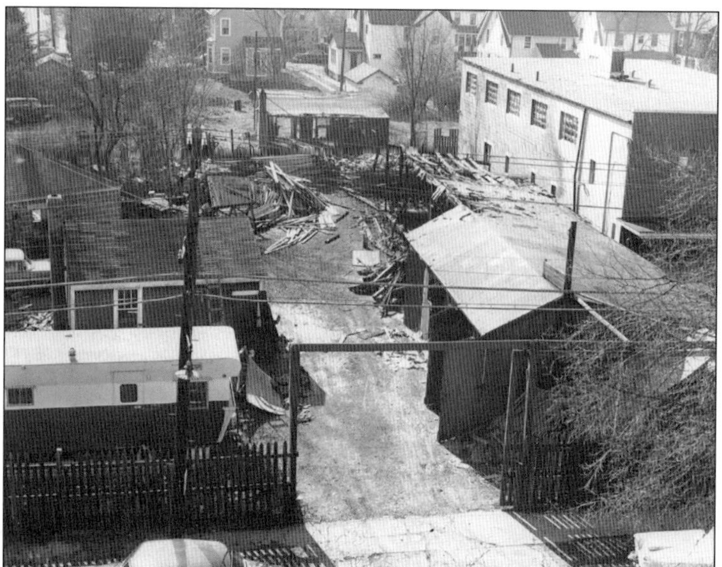

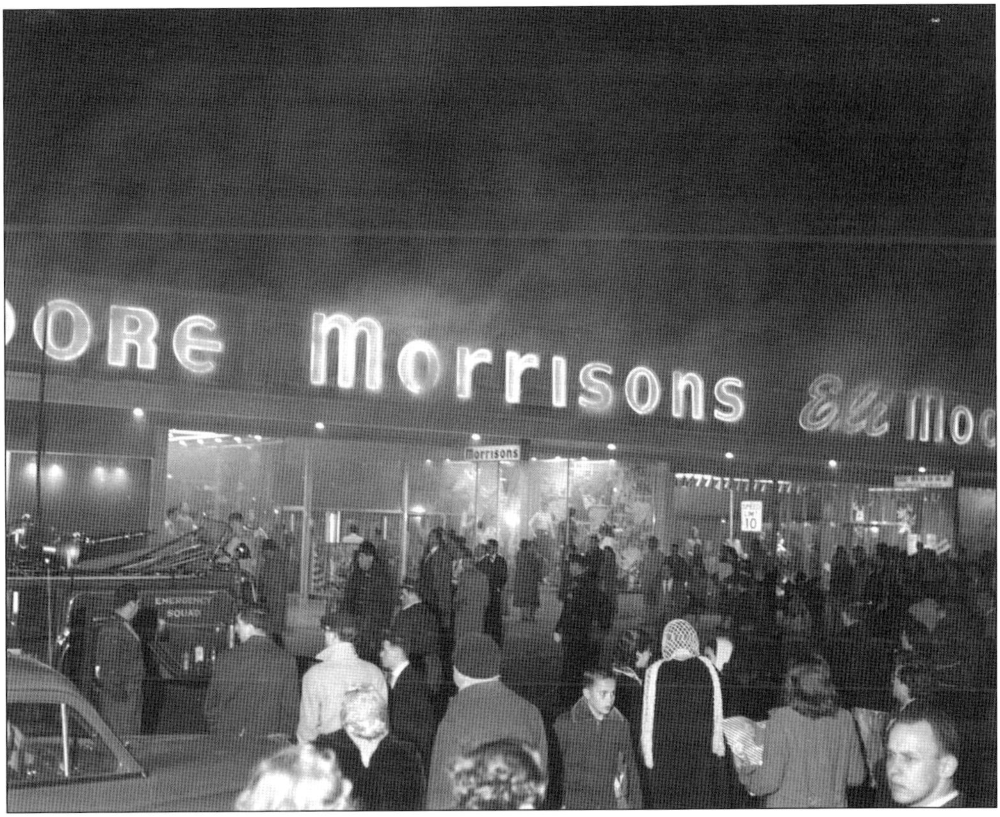

At about 5:25 p.m., just two days before Christmas 1957, a two-alarm fire of unknown origin started in a rear storage area of one of two Eli Moore department stores at the Hamden Plaza. The fire was out about a half hour after the first apparatus arrived. Losses were estimated at more than $100,000. (John Mongillo Sr.)

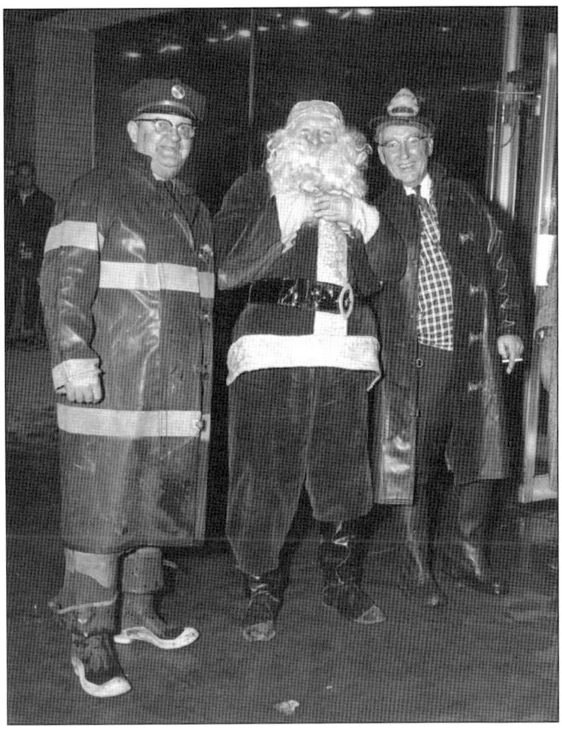

Once the Eli Moore fire was out, Battalion Chief Everett Doherty (left) and outgoing first selectman Herbert Hume flanked Santa, who was at his regular post in the Hamden Plaza to hear the kiddies' last-minute requests for Christmas 1957. Hume had just lost the recent municipal election to John DeNicola Sr., who would serve as Hamden's chief executive for five terms, with the last as Hamden's first mayor. (John Mongillo Sr.)

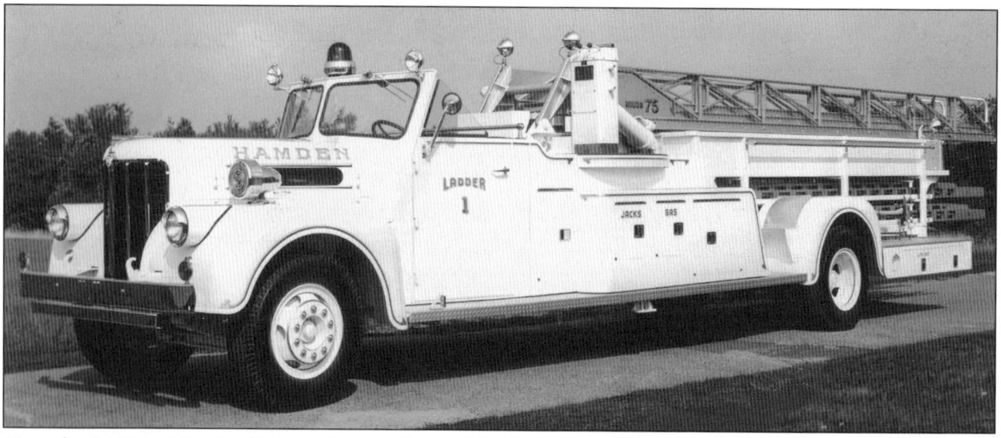

Hamden's 1958 Maxim 75-foot "Junior" aerial ladder truck was delivered in November 1958 and assigned to Humphrey Station 2. The Hamden Board of Fire Commissioners opted for white, as New Haven had done 15 years earlier. However, when a new town charter gave the chief more authority in making such decisions, Chief V. Paul Leddy, no fan of white apparatus, had this truck repainted red in 1971. (HFRA.)

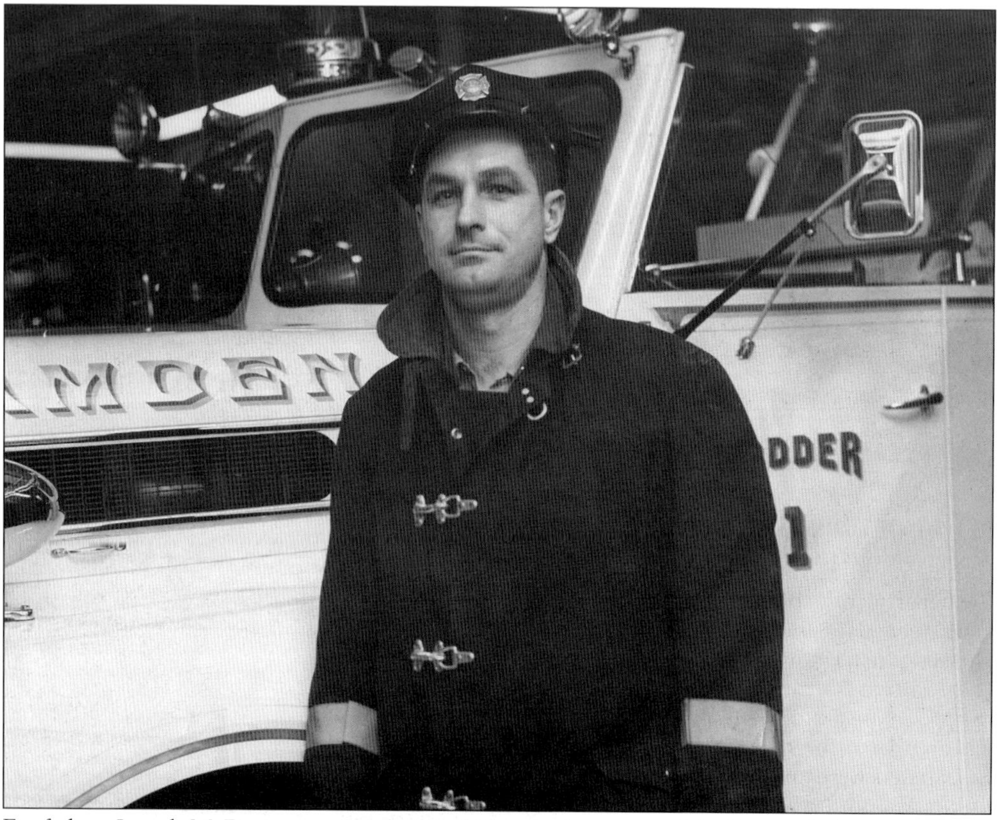

Firefighter Joseph McDermott was a driver of Hamden's new aerial ladder truck. Five years after this 1958 photograph was taken, McDermott was promoted to lieutenant when a second lieutenant position was added to each of the three platoons. Promoted to captain in 1970 and deputy chief (shift commander) in 1973, Deputy Chief McDermott, a "fireman's fireman," retired in September 1991. He is a member of the HFRA. (Lawrence Gershman.)

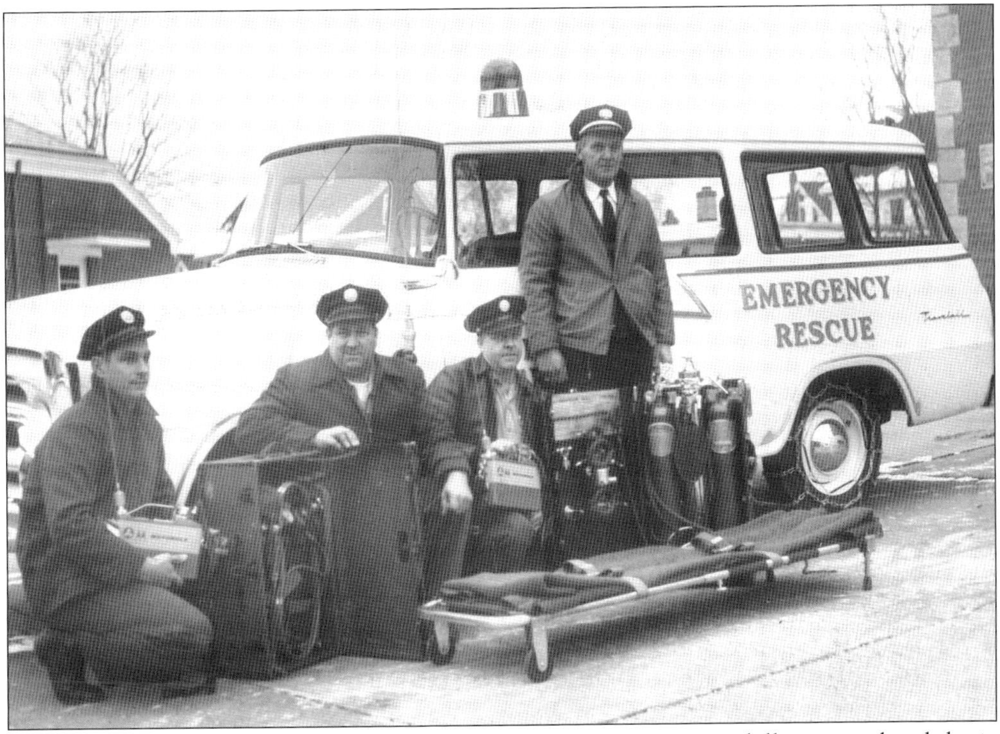

Hamden's first rescue unit was this 1958 International-Harvester Travelall, equipped with basic emergency first-aid equipment, including a stretcher, blankets, and the E and J Resuscitator. Pictured here on the front ramp at Humphrey Station 2 are, from left to right, Firefighters Joseph McDermott, Michael Cantarella, and David Herrmann and Capt. James Strain, whose father was the first captain of the Mt. Carmel Volunteer Fire Co. in 1911. (I.A. Sneiderman.)

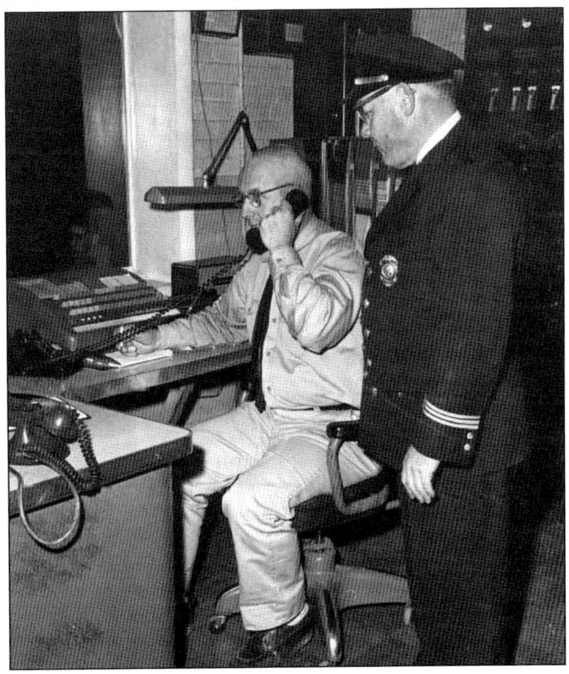

Firefighter Wilbur Baker (seated) was the department's chief dispatcher for many years. He is pictured here with Deputy Chief Training Officer Daniel Hume in the Alarm Room at the rear of Centerville Station 4, which was also department headquarters. Dispatchers were uniformed career firefighters until the late 1970s, when civilian dispatchers were hired under the Comprehensive Employment and Training Act (CETA), then made permanent employees. (Dominic Vaccaro.)

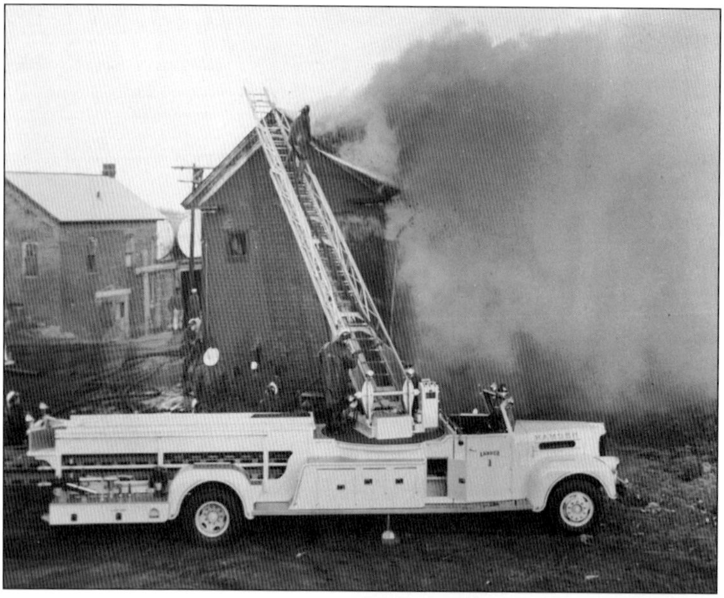

On January 31, 1959, Hamden's brand-new aerial ladder truck received its first workout at the North Haven Brick Yard on State Street in Hamden. The unsecured firefighter working the handline while on the aerial ladder is somewhat unconventional by today's standards, but it did provide some advantage in this situation. (Donald Yosua.)

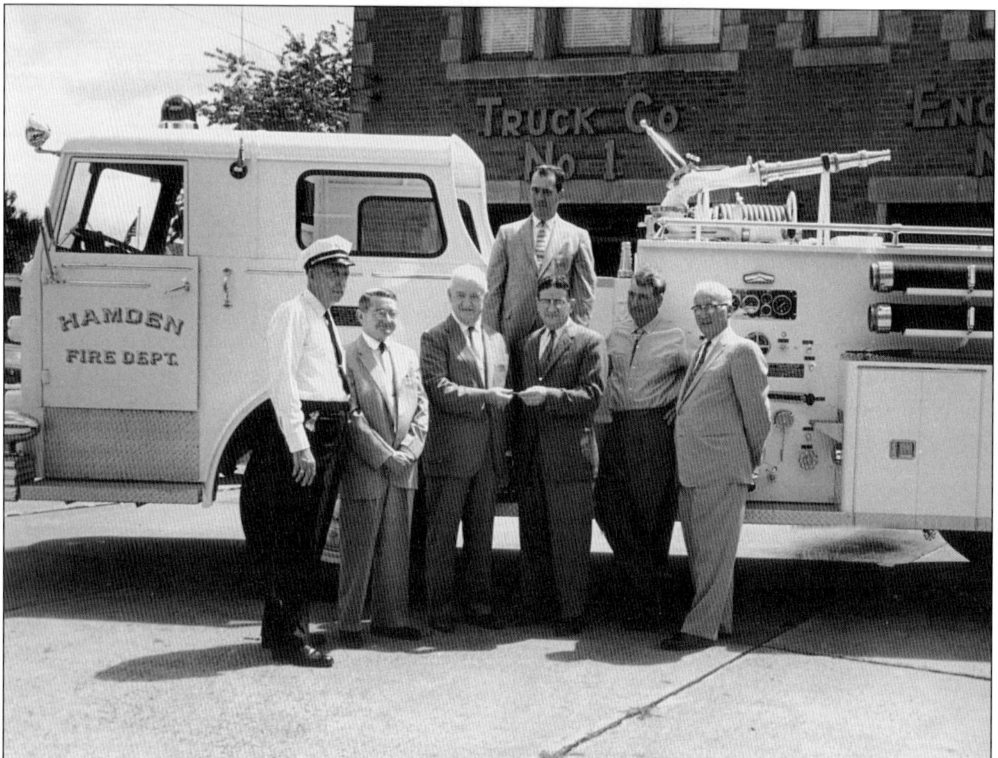

Accepting delivery of the new 1959 Maxim 750-gpm pumper at Humphrey Station 2 are, from left to right, Chief Raymond C. Spencer, Commissioner Michael Sullivan, Patrick Monahan of apparatus vendor O.B. Maxwell, first selectman John DeNicola Sr., commission chairman Elton Wetmore (on running board), Commissioner George Hindinger, and former first selectman Herbert Hume. One of Maxim's first cab-forward apparatus, this pumper is currently owned by three Hamden residents. (I.A. Sneiderman.)

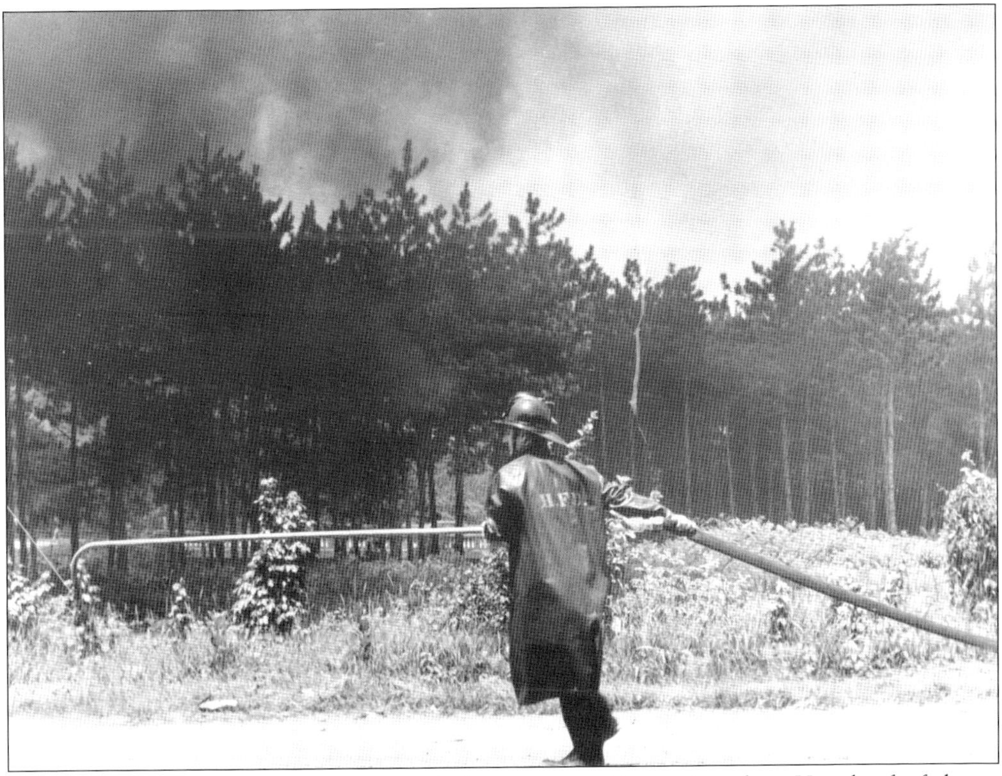

On July 8, 1959, disaster was averted through the courage and fast action of two Hamden firefighters. A tanker truck transporting 6,000 gallons of gasoline caught fire on Dixwell Avenue in North Haven, a few yards from the Hamden town line. Responding on Engine 1, Firefighters William Esposito (above) and Burton Hillocks approached the burning tanker truck with a charged two-and-a-half-inch handline. Using a Rockwood low-velocity fog applicator, they quickly extinguished the mass of flaming gasoline vapors impinging on the tanker. Chief Raymond Spencer (below) examines the damage to the cab. The following month, one firefighter was killed and 60 more injured in Kansas City, Kansas, when a gasoline storage tank exploded during firefighting operations, which introduced the term BLEVE (Boiling Liquid Expanding Vapor Explosion). This fire had the same potential. (Above, Donald Yosua; below, I.A. Sneiderman.)

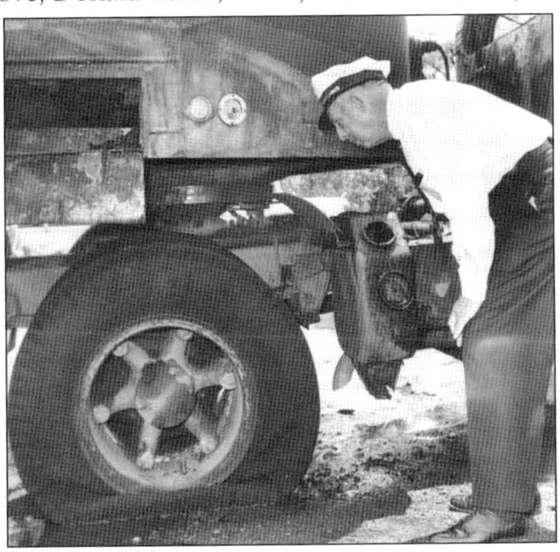

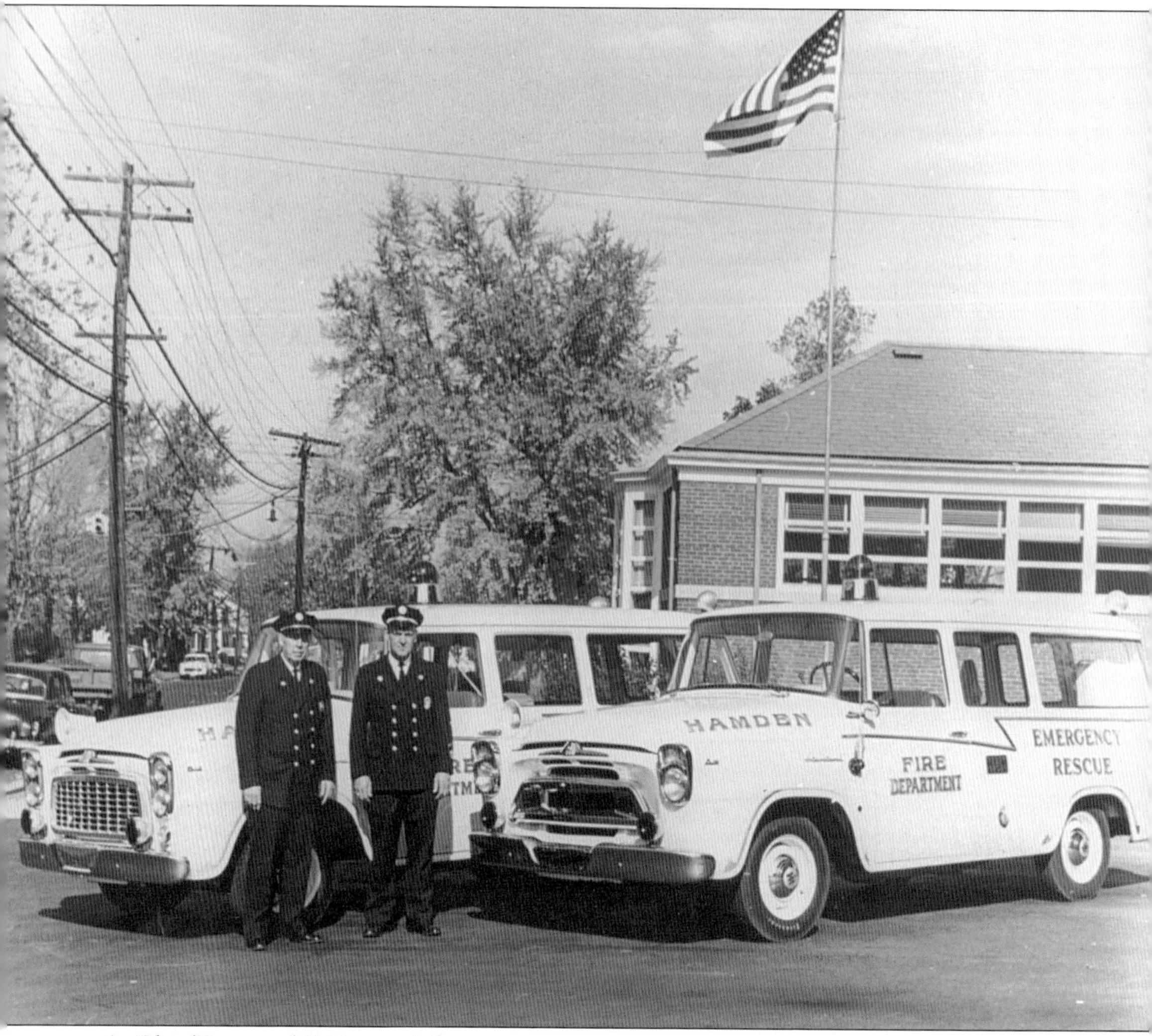

As Chief Raymond C. Spencer's fire service career was winding down in late 1960, he posed with Capt. James Strain (right) at Humphrey Station 2 with the department's two rescue units. This was when Hamden's emergency medical services consisted mainly of "oxygen calls." That would all change by the early 1970s with the introduction of the department's first emergency medical technicians. Rescue 2 (left) went in service at Centerville Station 4 on February 9, 1960. Rescue 1 was placed in service at the Humphrey station on December 10, 1958, and was later transferred to Whitneyville Station 3 on Putnam Avenue. The two International-Harvester Travelalls were in service until November 1971. (HFRA.)

Four

EXPANSION

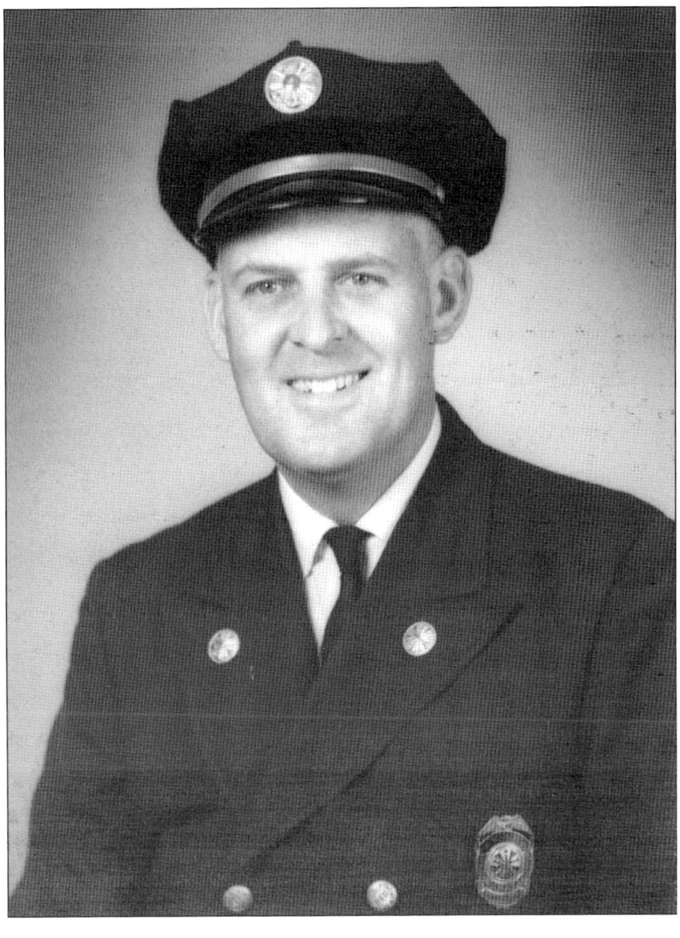

V. Paul Leddy joined the department in November 1941. Following four years in the Army during World War II, he returned to the department in 1946. Leddy was promoted to lieutenant in 1949 and captain in 1951. As a battalion chief, he became civil defense director in 1955 and was appointed chief in 1960. The department grew significantly during Chief Leddy's nearly 24-year tenure. (Fire Chief's Office.)

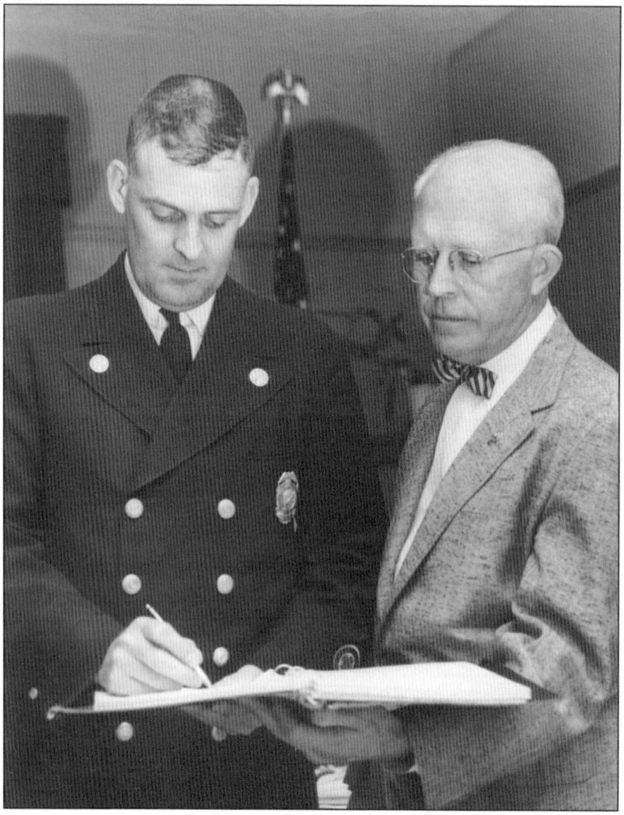

V. Paul Leddy is pictured here with town clerk Walter Connor after being sworn in as fire chief on November 17, 1960. He replaced Raymond C. Spencer, who was appointed Hamden's first paid fire chief in 1942. Chief Leddy would lead the department for nearly 24 more years. He passed away on April 29, 1994, almost 10 years to the day after he retired. (I.A. Sneiderman.)

In January 1961, Chief Leddy poses outside Humphrey Station 2 with Hamden's newest pumper and ladder truck. Leddy was not a fan of white fire apparatus. Hamden's next pumper and every other piece of apparatus thereafter would be red. It was no surprise that, 10 years after this photograph was taken, both trucks in this photograph were repainted red. (*New Haven Register.*)

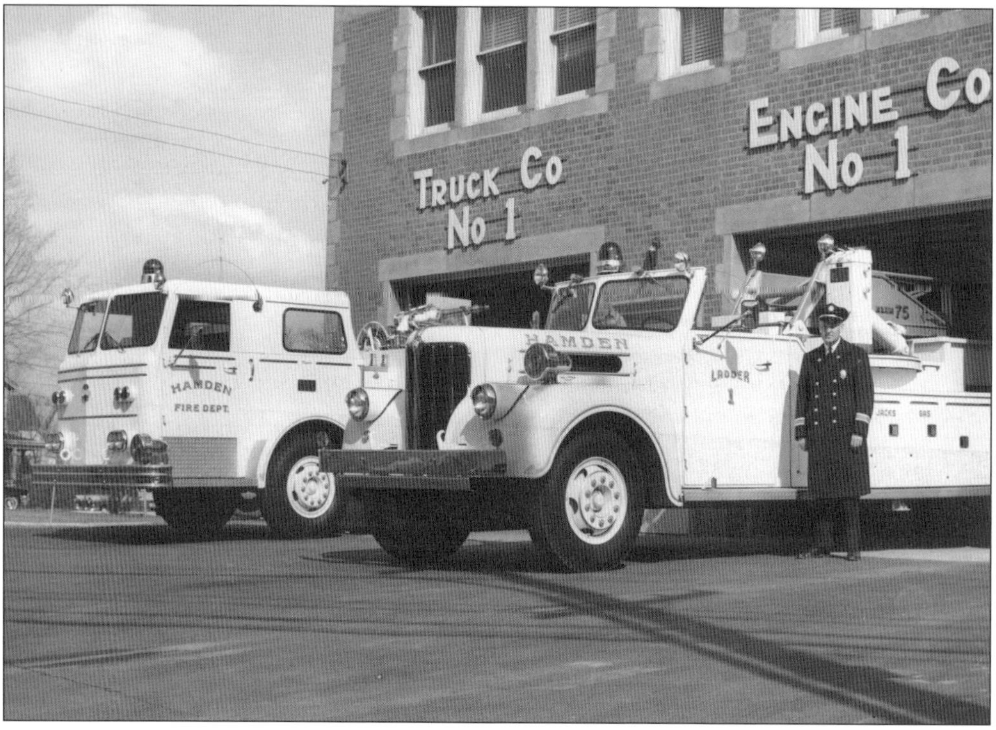

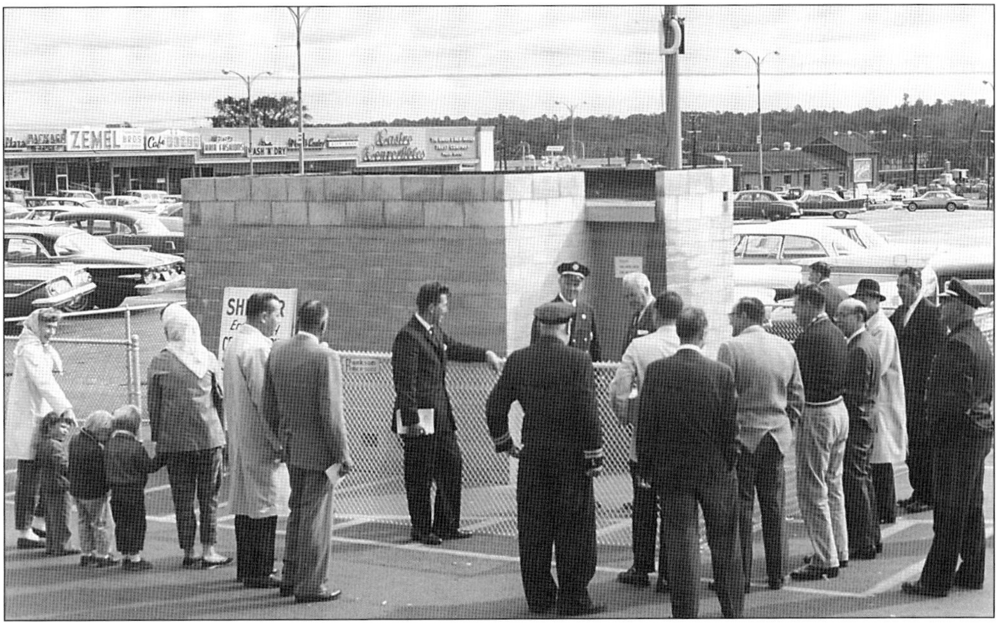

The unthinkable prospect of nuclear war was omnipresent during the mid-20th century. In 1961, a company selling backyard fallout shelters invited fire department leaders to inspect its wares at the Hamden Plaza. Chief V. Paul Leddy is inside the fence. Marshal Albert Purce's back is to the camera. The fire officer at the extreme right is Thomas Collins, chief of the New Haven Fire Department. (I.A. Sneiderman.)

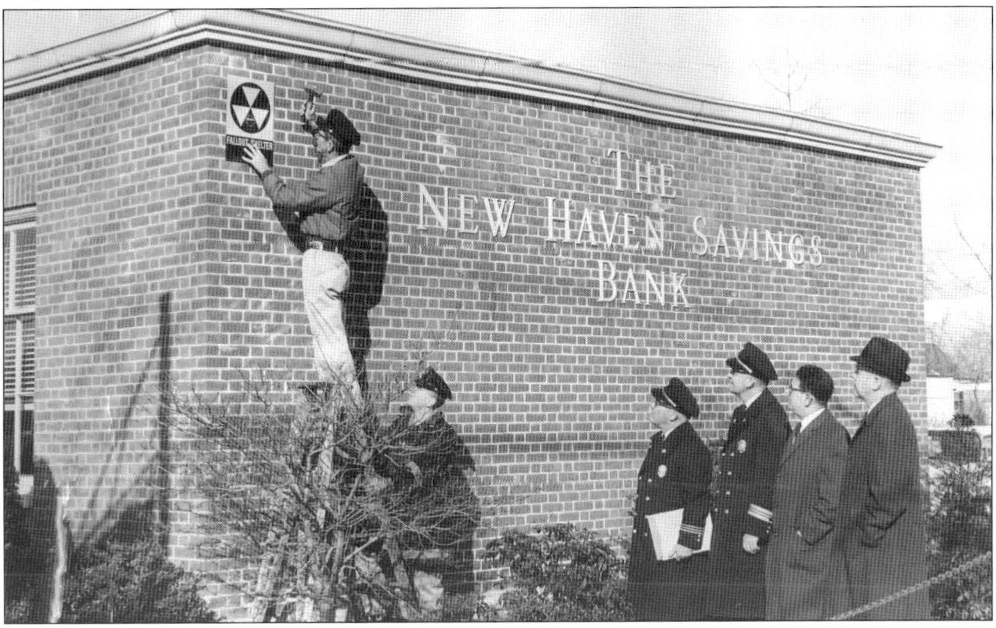

Civil Defense officials marked certain buildings that were offered as sanctuaries in the event of a nuclear attack. One such building was the New Haven Savings Bank in the so-called bank section of Dixwell Avenue. Department and other town officials look on as Firefighter Kenneth Harrington, on ladder, is assisted by Firefighter Richard Carney. In the 1970s, Harrington would become Hamden's Civil Defense director. (*New Haven Register.*)

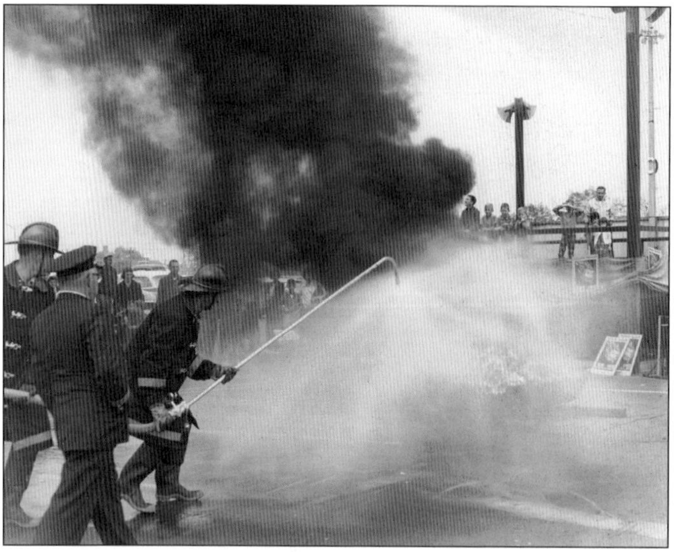

During Fire Prevention Week every October, personnel conducted demonstrations for the public at the Hamden Plaza shopping center. In this 1962 photograph, firefighters extinguish a kerosene fire in an open trough using an inch-and-a-half Rockwood nozzle and low-velocity fog applicator. (I.A. Sneiderman.)

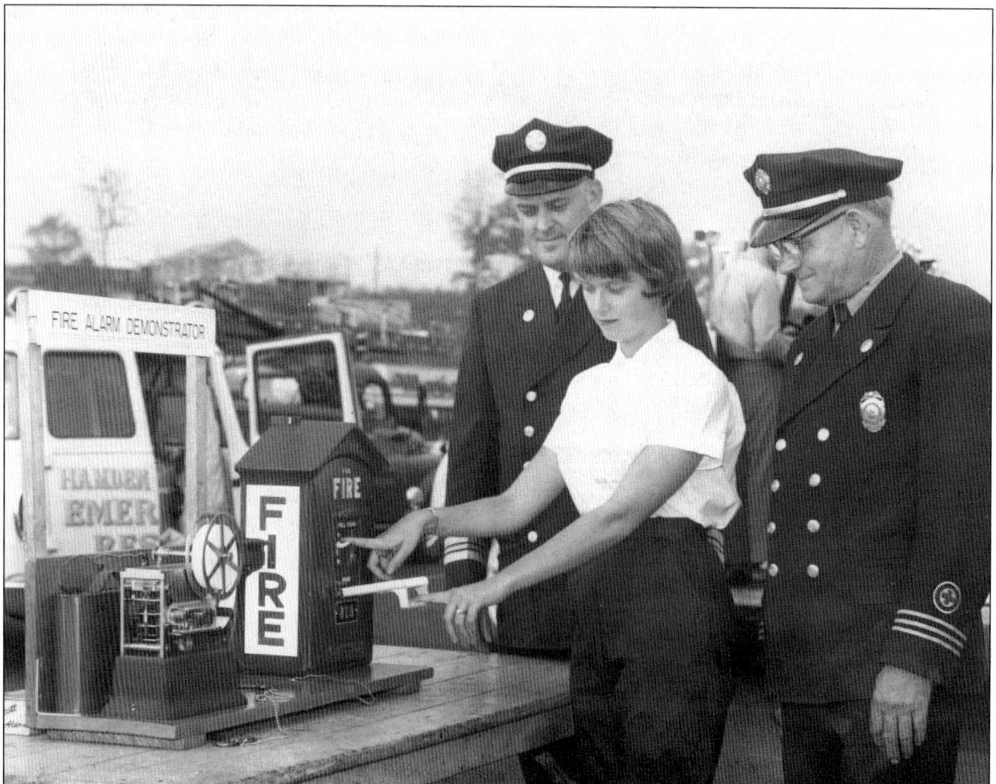

A popular Fire Prevention Week display was the Gamewell telegraph alarm system mock-up used to demonstrate how alarms are transmitted from utility pole–mounted alarm boxes to the fire stations. The young lady testing the alarm box in this 1962 photograph is flanked by Chief V. Paul Leddy (left) and Supt. of Alarms and Apparatus Clement Wetmore. (Domenic Vaccaro.)

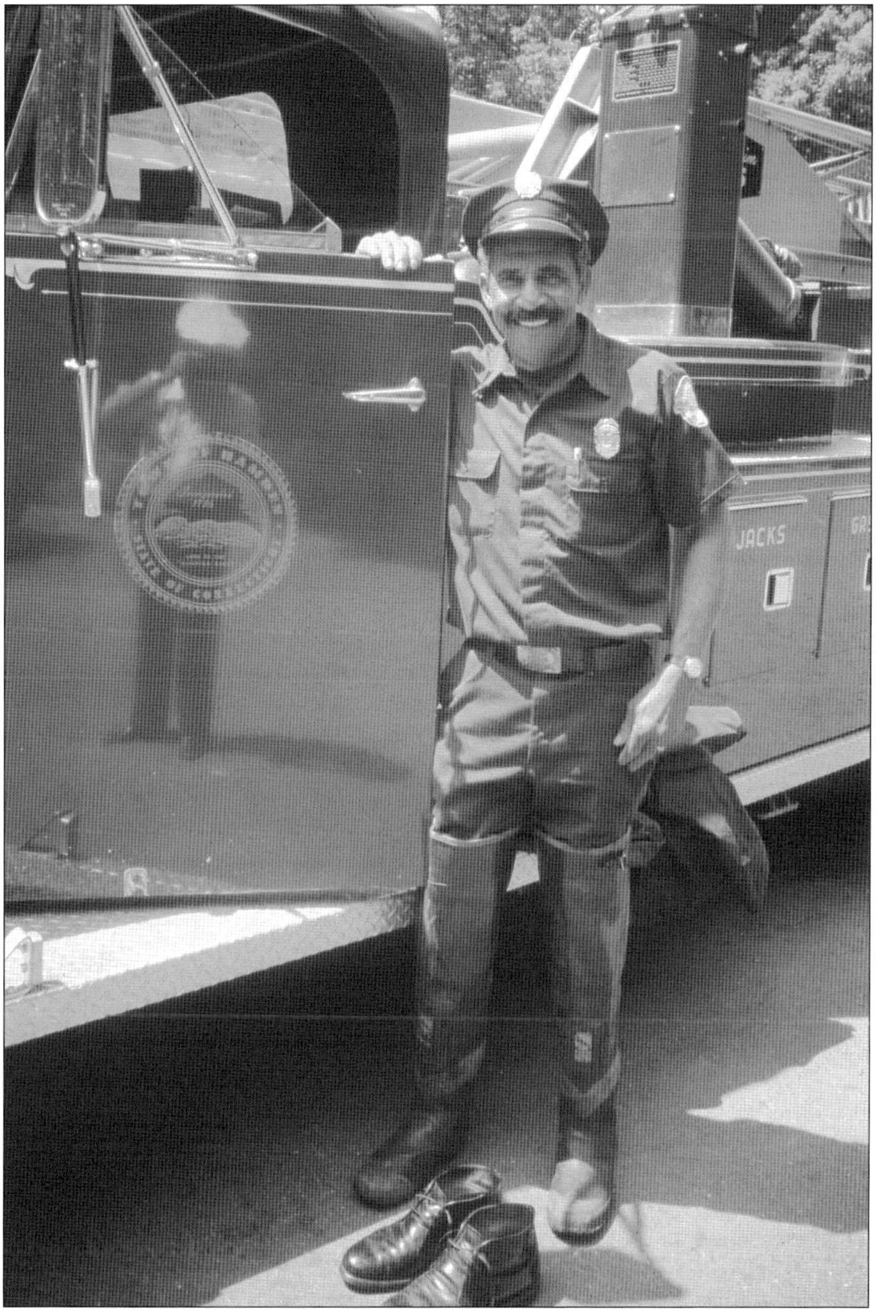

Then deputy chief John Tramontano took this 1982 photograph of Firefighter Samuel Augustus Jones standing beside Truck 1 during training at Co. 7. The department's first African American career firefighter, Jones joined the department 21 years earlier. Hailing from Knoxville, Tennessee, he also was the department's first four-year-college graduate. An excellent firefighter, Jones had an easygoing nature, sense of humor, and skill as an expert chess player that were legendary throughout the department. He retired in late 1990, after nearly 30 years on the job. On October 29, 2015, Samuel Augustus Jones passed away at his home in Knoxville, six days before his 90th birthday. His son Kenneth Jones is a career firefighter in Georgia. (Photograph by John Tramontano.)

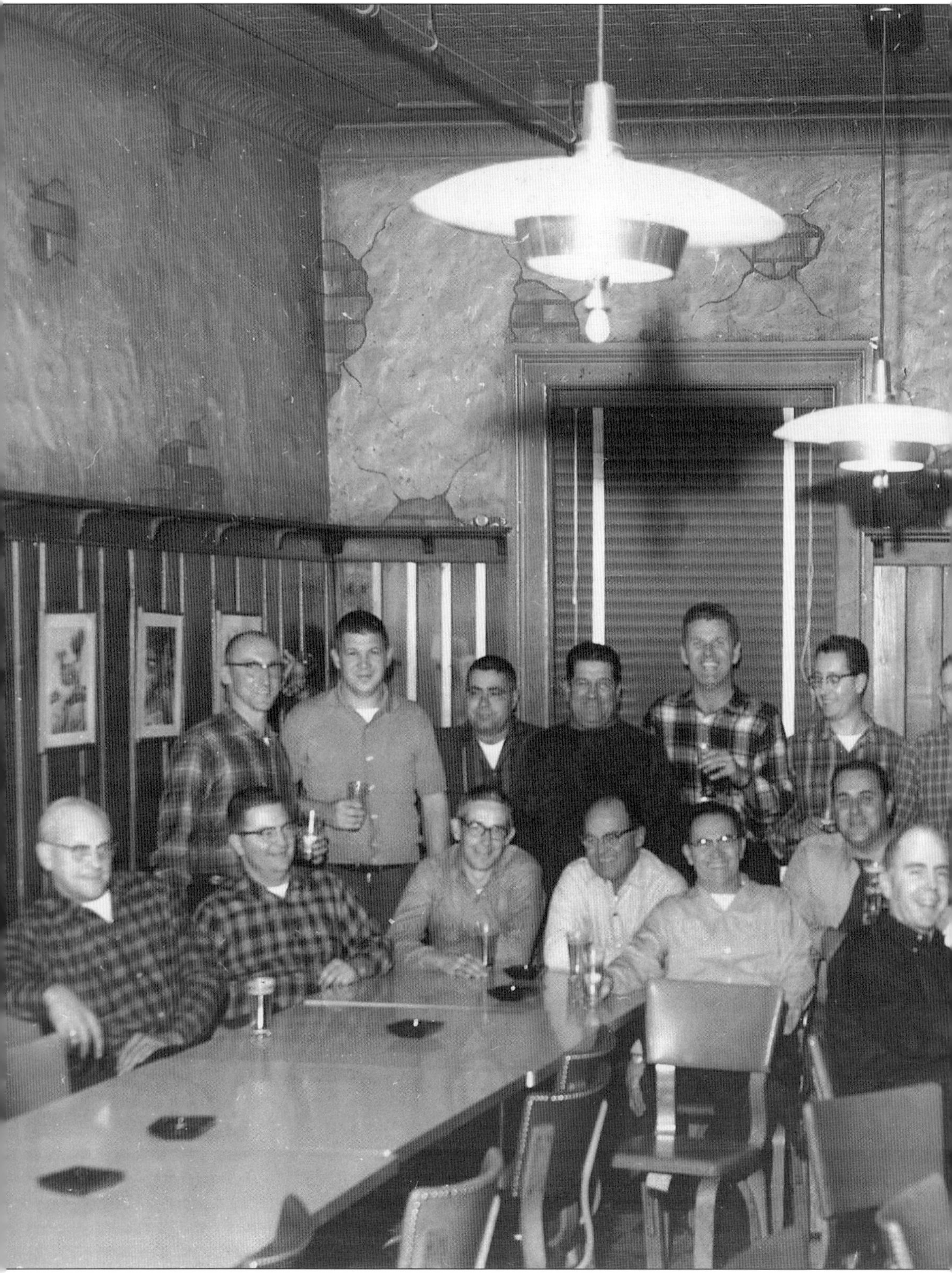

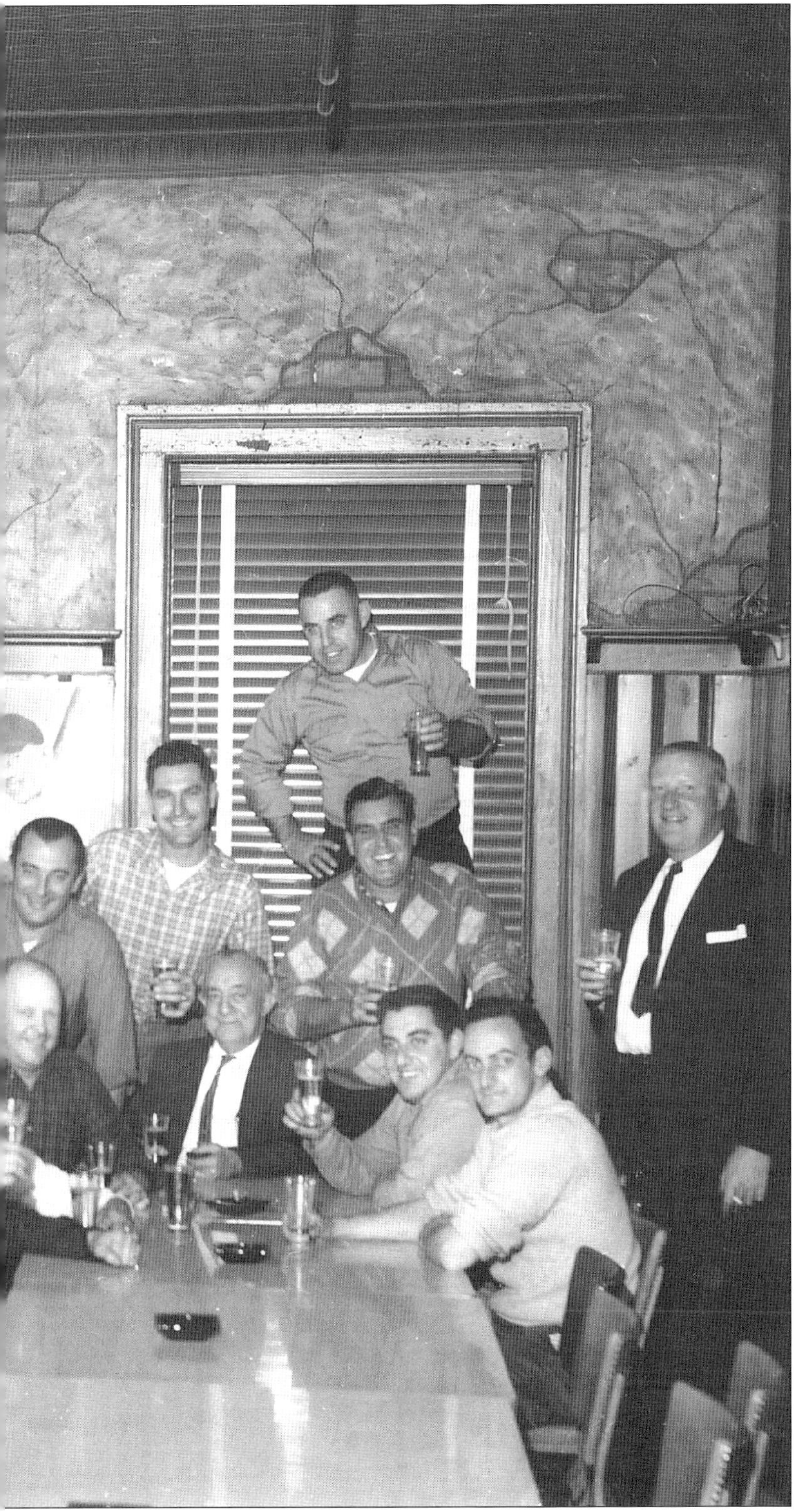

This photograph was taken in the rathskeller of New Haven's iconic Hull's Brewery on Congress Avenue in February 1962, when the members of Platoon 1 and a few guests got together for their "Semi-Annual Beer Party." From left to right are (seated) Emil Strain, Lawrence Bellemore, Francis Leddy, Joseph Hromadka, two unidentified men, Richard Carney, Joseph Rahl, David Herrmann, Clement Kammerer, Carmen Amarante, and Thomas Doherty; (standing) Warren Blake, Raymond Bantz, Salvatore "Cookie" Ferraro, Michael "Mickey" Cantarella, Hugh McLean, Gerald Wolf, John McKee, Stanley Brown, Joseph McDermott, Vincent Roth, Richard Stacey (on chair behind Roth), and Luke Tobin. (Thomas Doherty.)

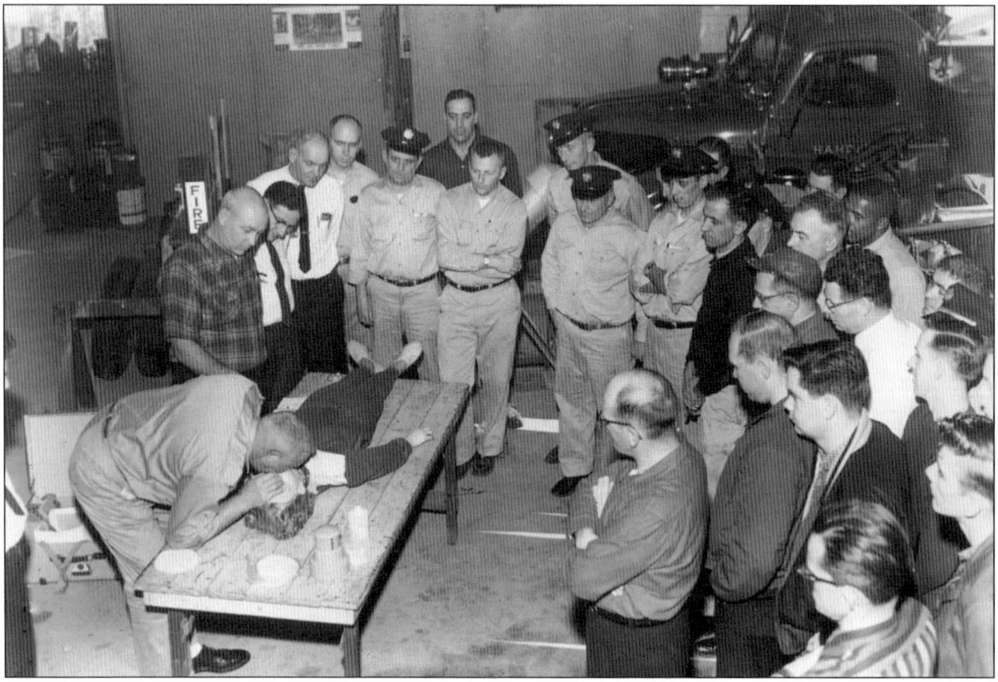

Many on- and off-duty, career, and volunteer personnel assembled at the repair shop for this 1962 mouth-to-mouth resuscitation training session. Firefighter Luke Tobin is pictured demonstrating the procedure. Among the participants was Hamden first selectman John DeNicola Sr. (second from left), who would become Hamden's first mayor in 1966. (Richard Carney.)

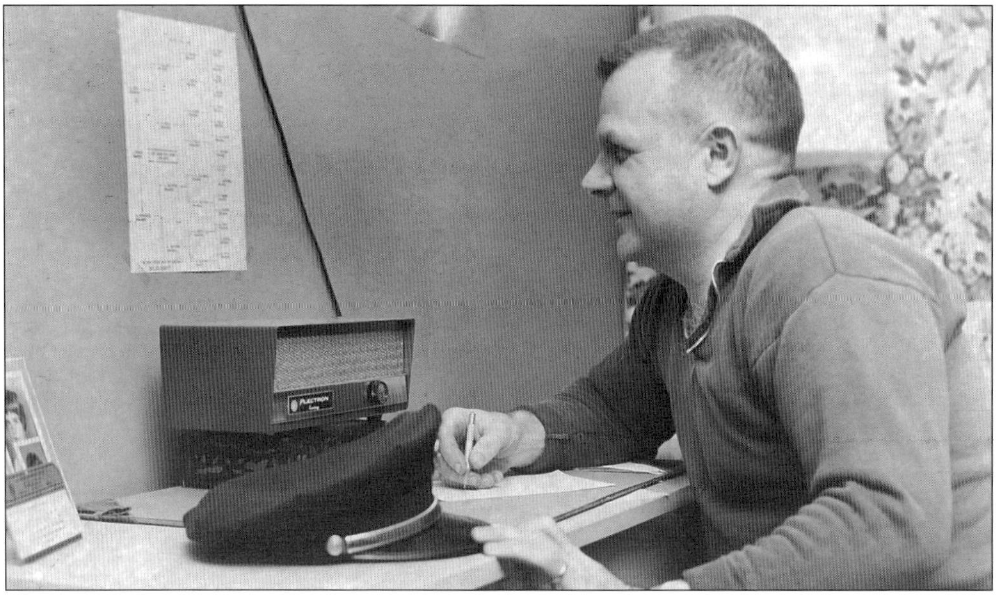

The 1963 sale of the 1941 Diamond-T ladder truck to New Milford financed the purchase of Plectron radios to alert Hamden volunteer firefighters. The radios were issued to the most active volunteers. Each company had its own tone alert. The dispatcher tested the receivers every day, activating the tones for all four volunteer fire companies at 1815 hours. Obsolete today, Plectrons were a blessing before pagers and cell phones. (I.A. Sneiderman.)

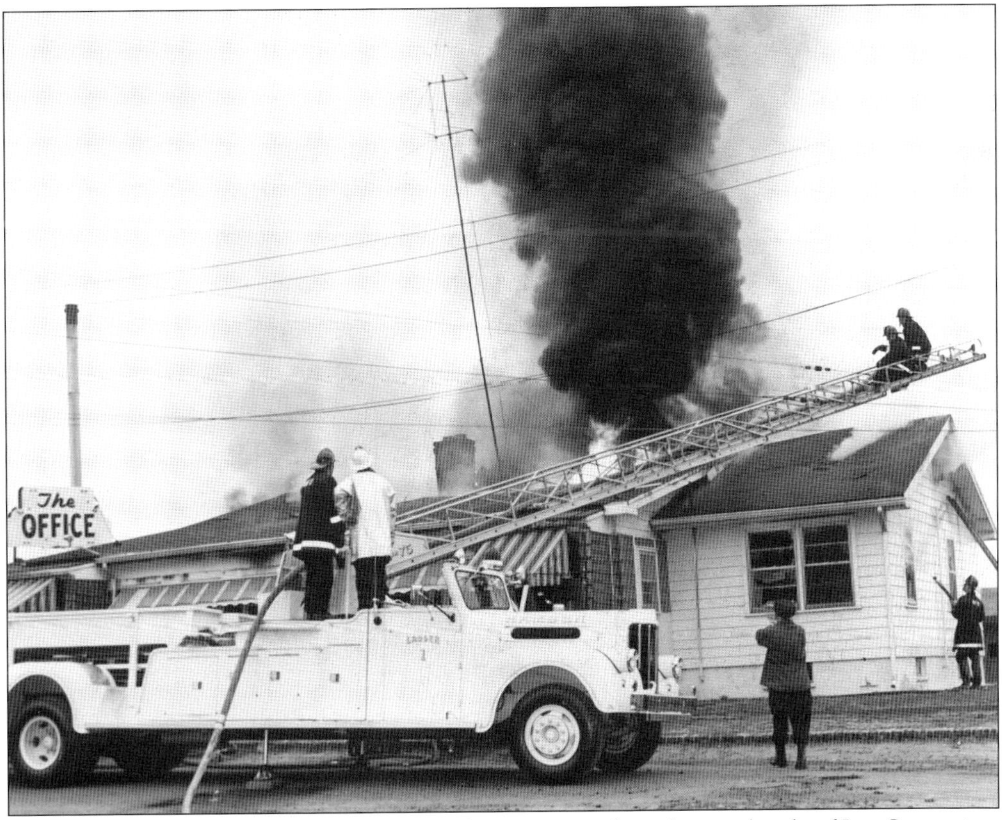

The well-known joke about this nightclub that once stood on the north side of Lee Street, just off Dixwell Avenue, was that a guy could call his wife at home and say that he would be late for dinner because he was "still at the Office." On the morning of April 3, 1963, the Office Restaurant, formerly Club Del-Mar, was destroyed by fire. (Domenic Vaccaro.)

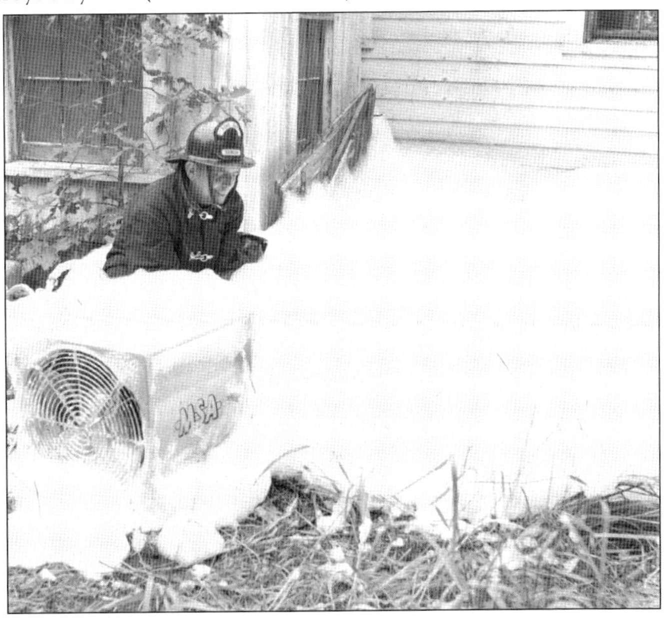

In August 1963, the department began experimenting with high-expansion foam to fight basement fires. Firefighter Robert "Ace" Callahan is pictured at the hatchway of a derelict building in the old Winchester "powder farm," off Putnam Avenue. The Olin-Mathieson Company, owners of the powder farm, donated several condemned buildings located there for live fire training. (HFRA.)

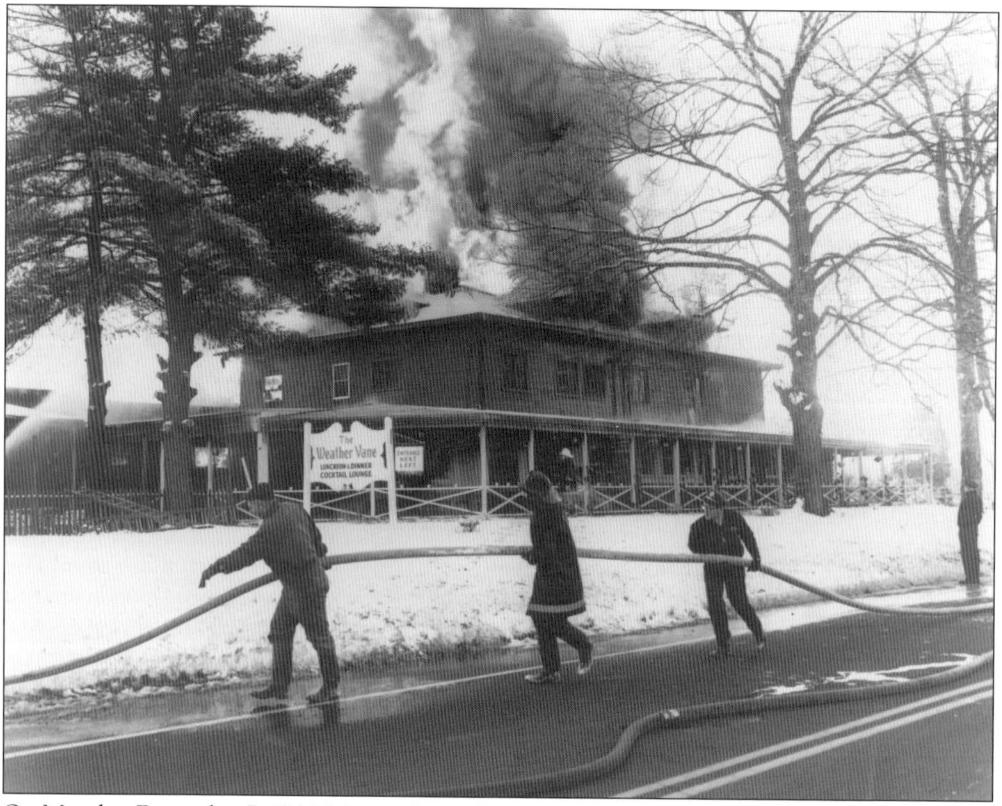

On Monday, December 7, 1964, Mount Carmel's popular Weather Vane Restaurant was destroyed in this general alarm fire of unknown origin. As apparatus were positioning, this unidentified firefighter (center) got an assist from a civilian and ambulance driver John Murray (right). The fire spread so quickly before being noticed that the owner and her two young daughters barely made it outside before flames consumed everything. (John Mongillo Sr.)

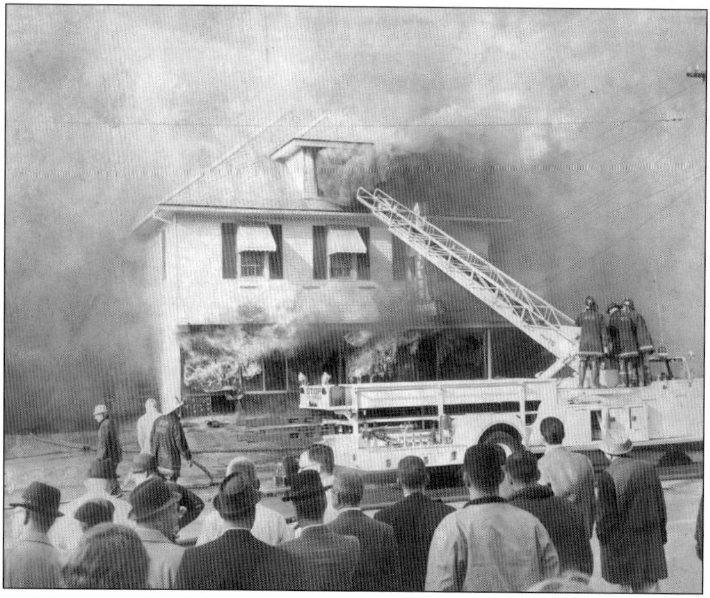

Less than three weeks after the Weather Vane burned, fire gutted Reilly's Restaurant on a very balmy Christmas Day. According to the *Hamden Chronicle*, John Mongillo Jr. observed smoke coming from the building and alerted beat cop Ronald Durkin. Mongillo, who would later become a noted news photographer like his father, ran to his nearby home to retrieve his camera, getting this shot and many others. (Photograph by John Mongillo Jr.)

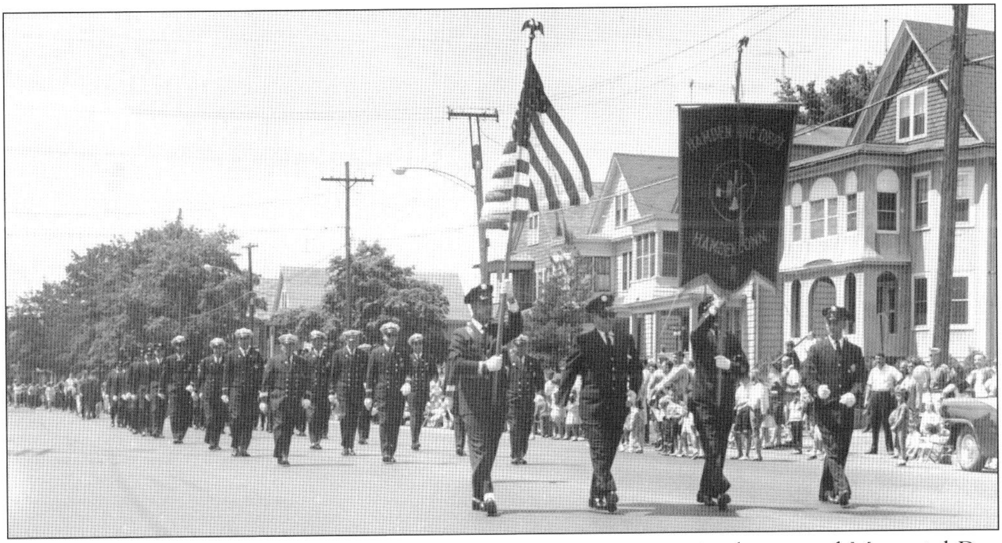

The Hamden Fire Department is always a prominent participant in the annual Memorial Day parade. For many years, marchers and fire apparatus assembled in Highwood and marched north on Dixwell Avenue to the high school, which was the route when this photograph was taken in 1964. The Honor Guard members are, from left to right, Firefighters Eugene Maturo, David Herrmann, Robert "Ace" Callahan, and Benjamin Mikolinski. (*New Haven Register.*)

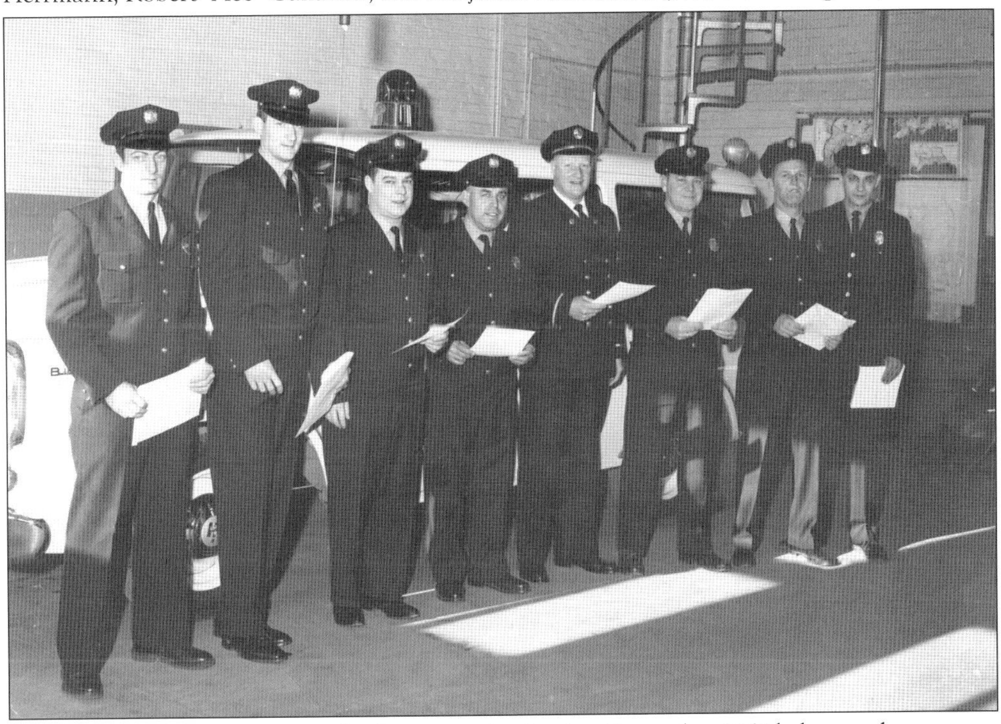

Receiving department citations for their participation in a December 1964 below-grade rescue on Skiff Street are, from left to right, Firefighters Thomas Doherty, Harold Mangler, Charles Carlson, and Richard Stacey; Lt. Luke Tobin; and Firefighters Burton Hillocks, Hugh McLean, and Edward Doiron. Carlson, Mangler, and Doiron had been in recruit training at the time. Mangler would serve a remarkable 41 years, 11 months with the department. (John Mongillo Sr.)

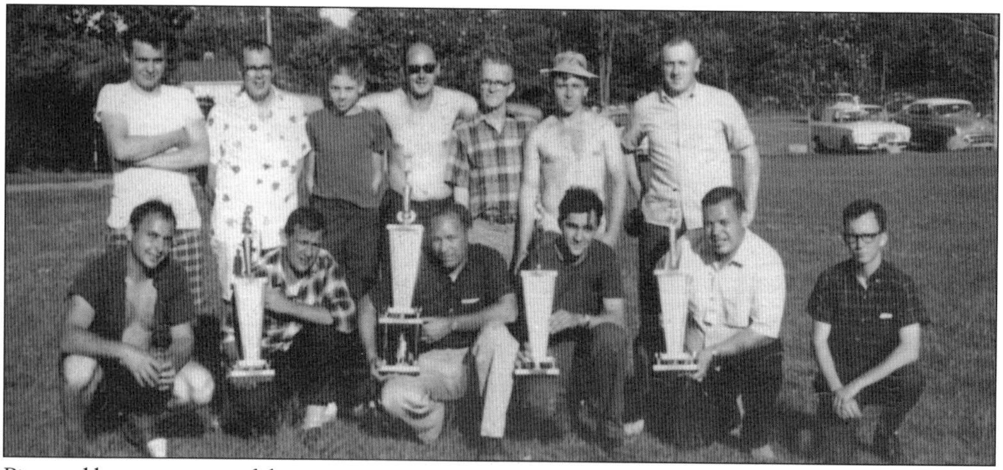

Pictured here are some of the winning team members in the first annual Hamden Fire Department Field Day held Sunday, June 14, 1964. From left to right are (first row) Robert Feinn, unidentified, Find Pedersen, Edward Doiron, Anders Pedersen, and James Mathis; (second row) Robert Ford, G. Donald Steele, Timothy Steele, unidentified, Toby Milum, William Scott, and Richard Pleines. (HFRA.)

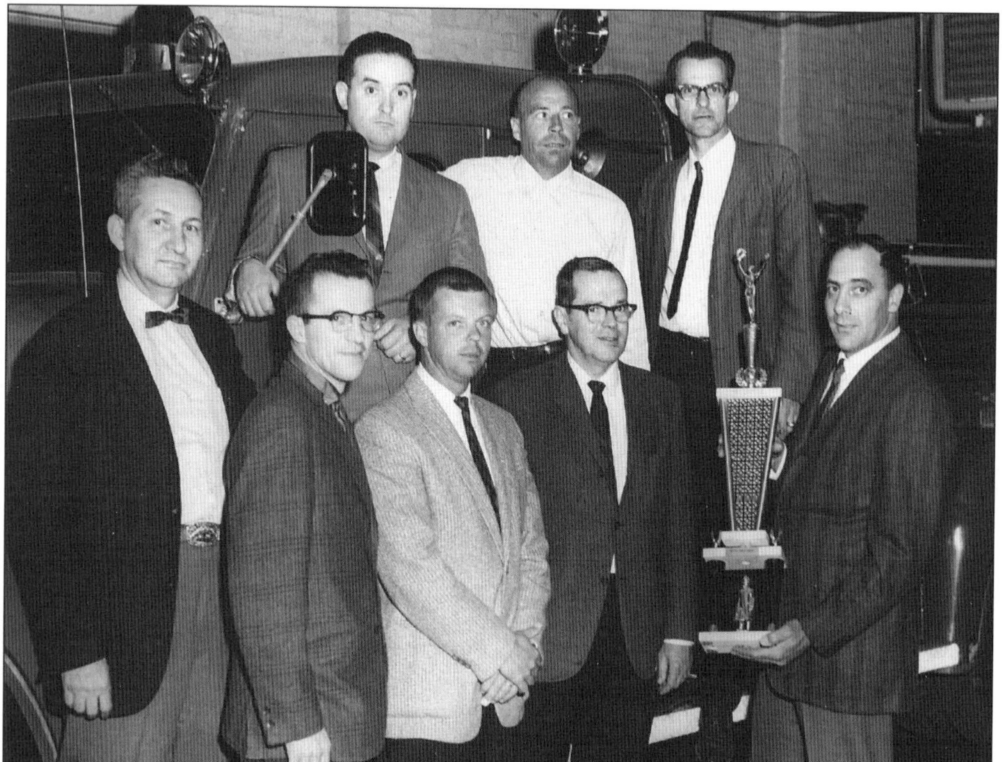

Volunteer members met to plan the second annual Hamden Fire Department Field Day, which took place on the second Sunday of June 1965. From left to right are (first row) Robert Smith (Co. 9), Duane Wetmore (Co. 7), Leonard Sablitz (Co. 9), G. Donald Steele (Co. 5), and Norris Buchter (Co. 8); (second row) Joseph Cuomo (Co. 9), Kenneth Johnson (Co. 8), and Kenneth Vowinkle (Co. 7). (*New Haven Register.*)

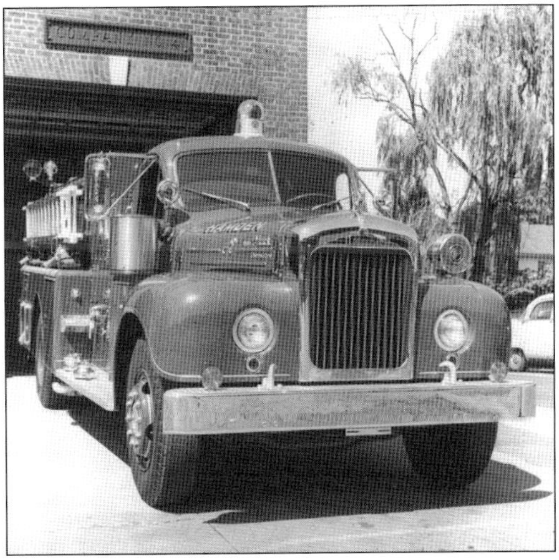

This 1965 Mack 750-gpm pumper was the first new pumper purchased after the department returned to red apparatus. It was placed in service temporarily as Engine 2 on August 18, 1965 while the 1959 cab-forward was being repaired, then permanently assigned to Centerville Station 4 as Engine 4. The Mack was Hamden's first new full-size apparatus on a commercial chassis since the Diamond-Ts purchased before World War II. (HFRA.)

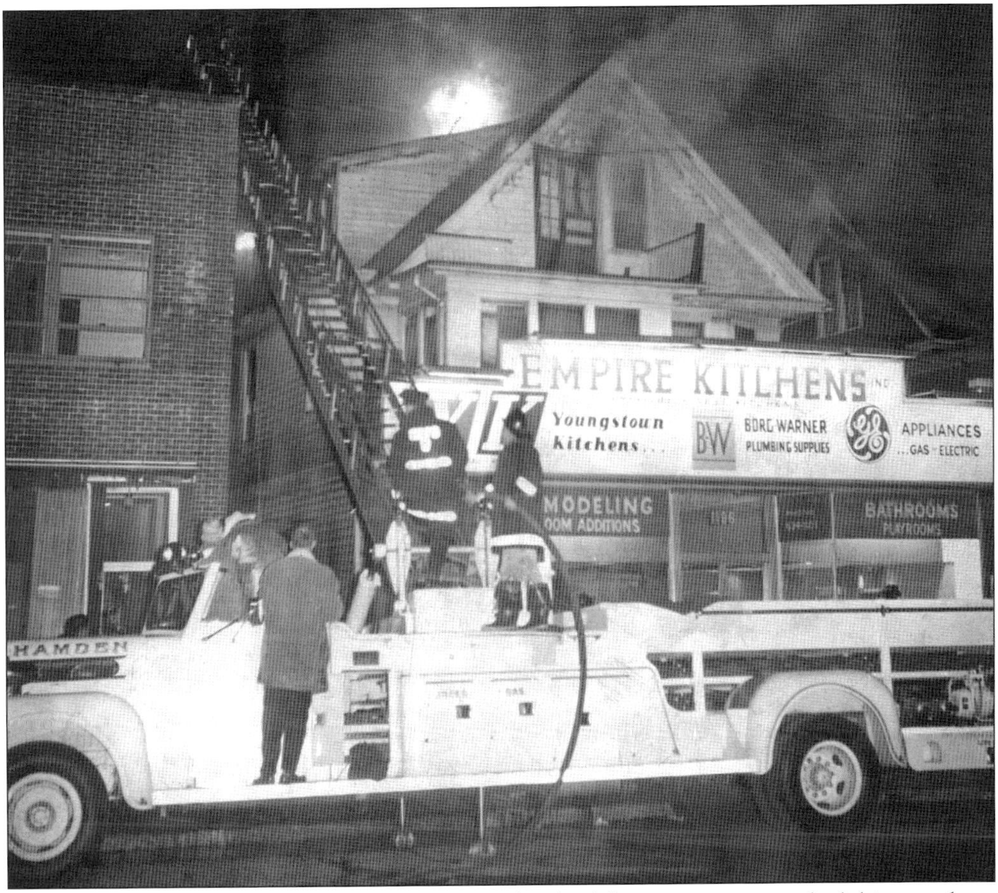

According to newspaper accounts, a gas heater in the third-floor apartment sparked this two-alarm fire on the second to last day of 1965. The third-floor occupant went to turn on the heater shortly after 7:00 p.m., found it enveloped in flames, and fled the building. Fortunately, the second-floor apartment occupants were away. The first floor housed Empire Kitchens, Inc., which suffered extensive water and smoke damage. (HFRA.)

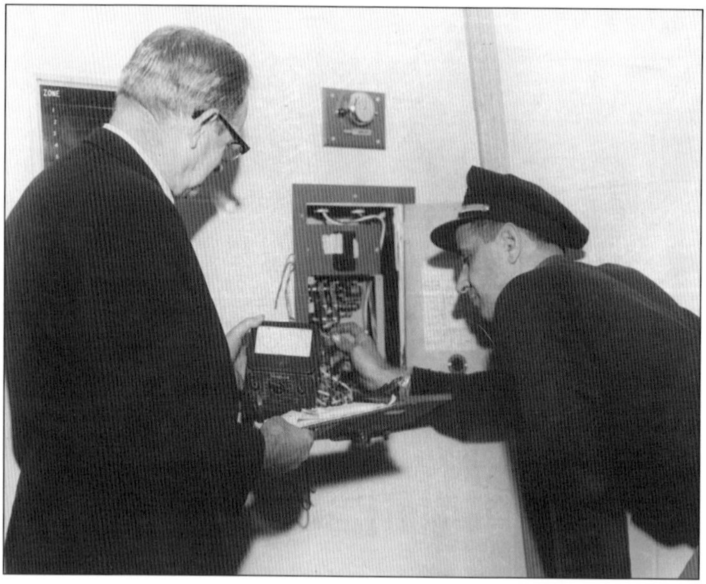

In this late-1960s photograph, Supt. of Alarms and Apparatus Clement Wetmore (left) and Asst. Supt. Richard Lostritto are testing the internal alarm circuit at a Hamden elementary school. The alarm circuit, or "loop," was connected to the school's Gamewell master alarm box that, when tripped, sent the code for that school to all fire stations. (HFRA.)

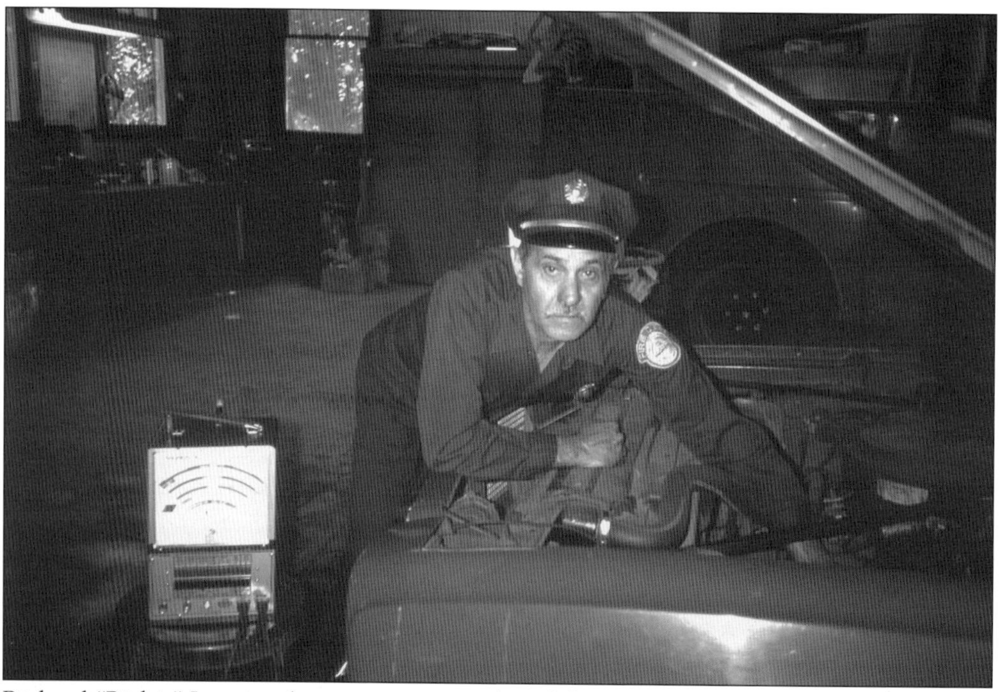

Richard "Richie" Lostritto became superintendent following the 1968 retirement of Clement Wetmore. In addition to maintaining the department's vast alarm system, Lostritto and Asst. Supt. Paul Wetmore Sr. also maintained the department's numerous staff cars and 15 pieces of career and volunteer fire apparatus. Always a busy man, Lostritto was photographed here in 1982 tuning up the 1974 Chevy Malibu wagon that was once Car 30. (John Tramontano.)

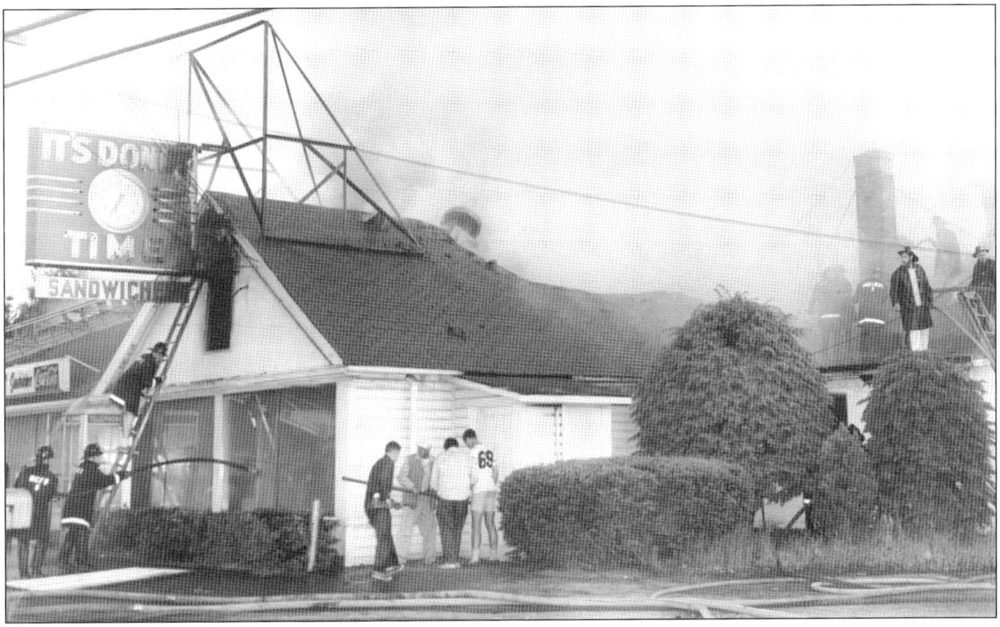

The final event of the department's third annual Firemen's Field Day was not on the schedule at Brooksvale Park on June 12, 1966, when the competition was cut short by a fire that destroyed the Connecticut Doughnut Company at 3800 Whitney Avenue. This may explain why some of the firefighters are rather informally clad beneath their turnout gear or have no gear at all. (John Mongillo Jr.)

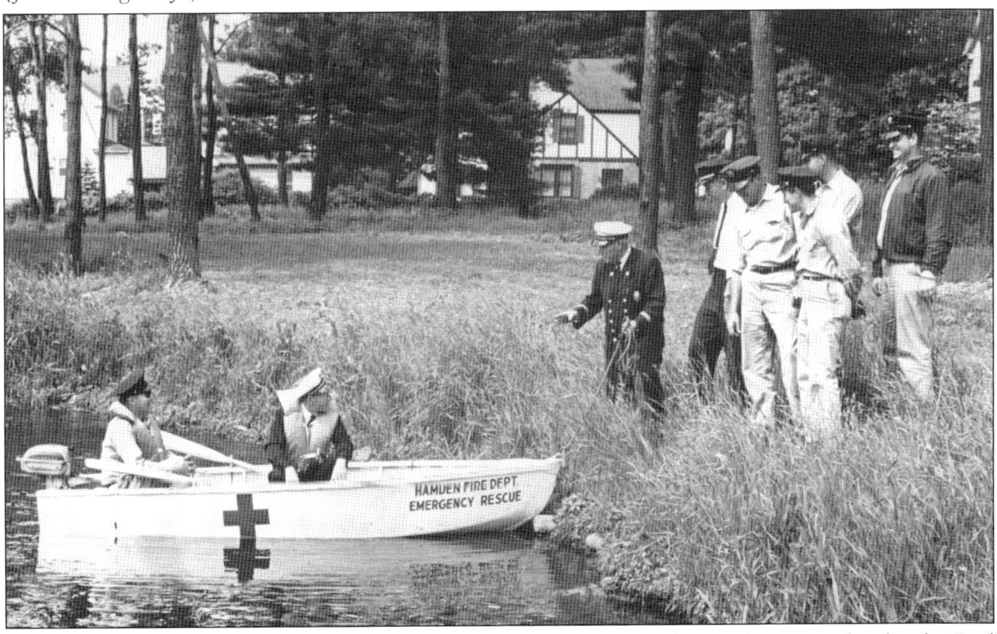

Firefighters trained every spring at Lake Whitney with the boat donated years earlier by the Red Cross. Firefighter Milner Benham (left) and Lt. William Hines are manning the boat in this 1966 photograph. Those on land are, from left to right, Deputy Chief Training Officer Daniel Hume, Lt. Daniel O'Connell, and Firefighters Arthur Heriot, Paul Wetmore Sr., Paul Reutenauer, and Gilbert Spencer. (*New Haven Register.*)

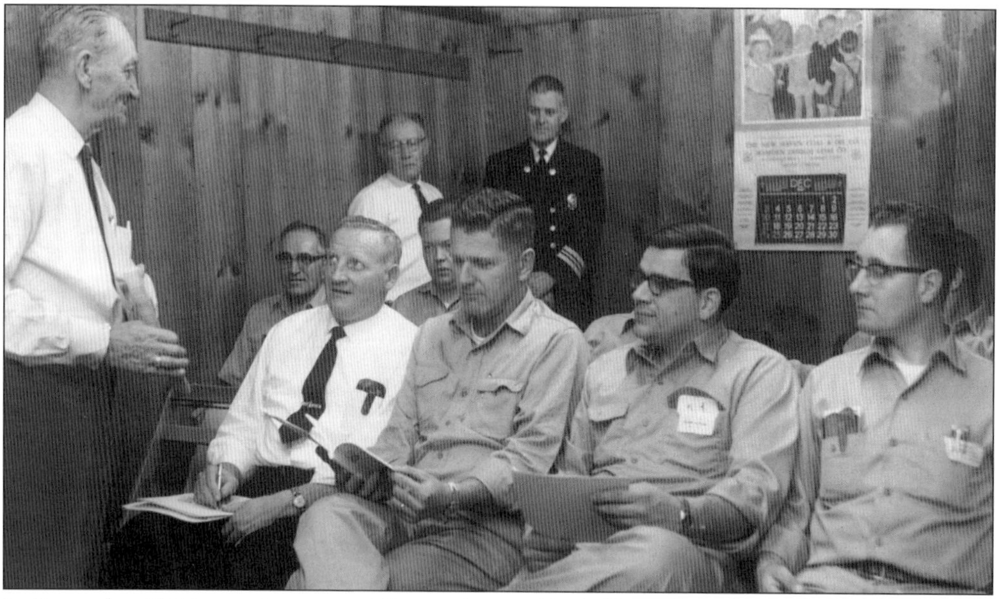

In December 1967, state instructor Thomas Forbes (far left) conducted civil defense training at Centerville Station 4. From left to right are (first row, seated) Lt. Luke Tobin, and Firefighters Hugh McLean, William LaVelle, and Gerald Wolf; (second row, seated) Firefighters Mario "Bucky" Serafino and William Davin. Standing in the rear are Deputy Chief Training Officer Daniel Hume (left) and Chief V. Paul Leddy. (Domenic Vaccaro.)

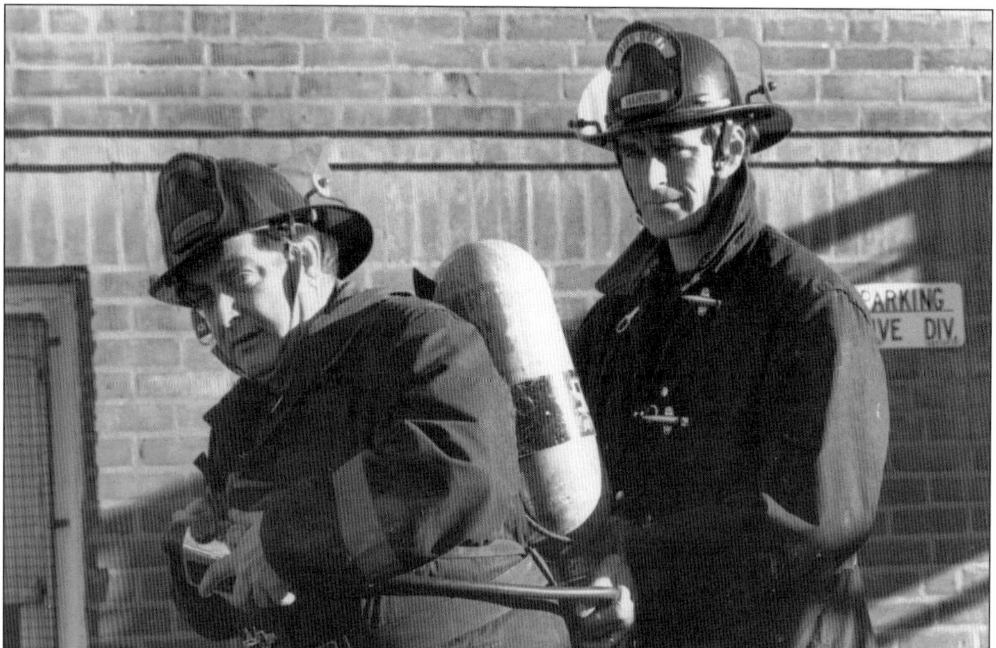

This undated photograph of Firefighters Gaetano "Carmen" Amarante (left) and Paul Reutenauer taken behind Centerville Station 4 appears to be a demonstration of the positive pressure Scott 2A self-contained breathing apparatus (SCBA), which was state-of-the-art at the time this photograph was taken. (HFRA.)

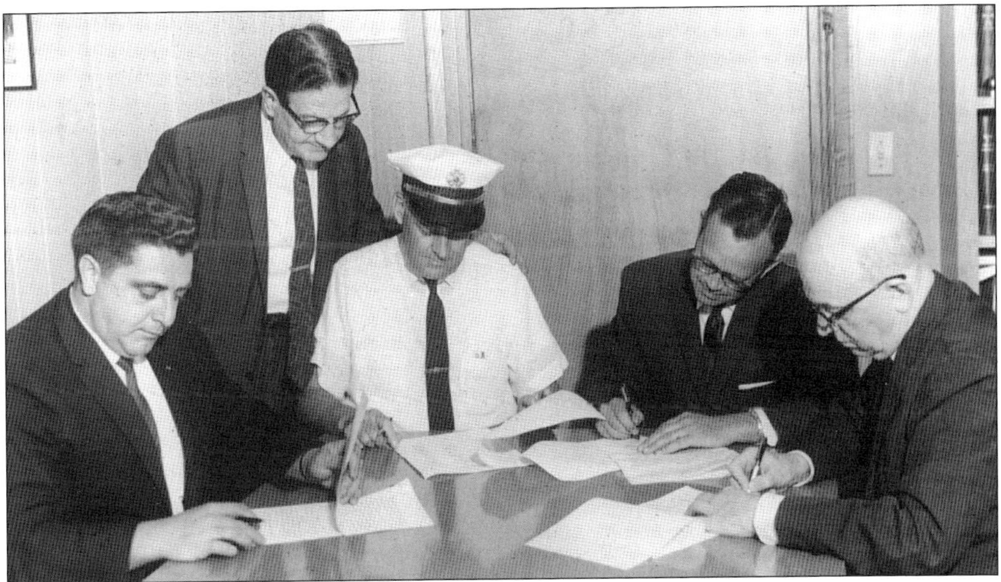

After spirited negotiations and a town-wide referendum affirmation, the town and the Hamden Paid Firemen's Sick Benefit Association signed an agreement for a 42-hour workweek to commence in October 1970. Pictured here in August 1967 are Mayor John DeNicola Sr., standing, and, from left to right, Firefighter Carmen Amarante, Chief V. Paul Leddy, Firefighter Frederick Knudsen (association president), and an unidentified town representative. (HFRA.)

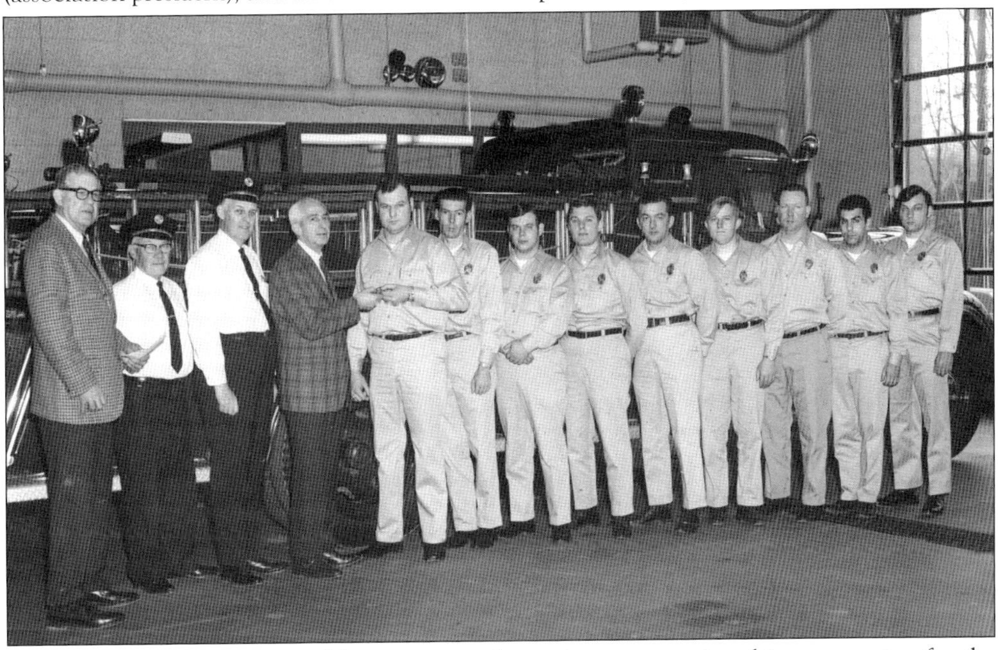

During 1969 and 1970, several large groups of recruits were appointed in preparation for the upcoming 42-hour workweek. In this January 1970 photograph are, from left to right, Commissioner Irving Saslow, Deputy Chief Training Officer Daniel Hume, Chief V. Paul Leddy, Commissioner Ted Baroncini, and Firefighters Ralph Dievert, Ernest Braun, Raymond Ramelli, Ralph Tortora, Joseph Mulligan, William Marak, John Reynolds, Edward DeFrancesco, and Raymond Dobbs. (I.A. Sneiderman.)

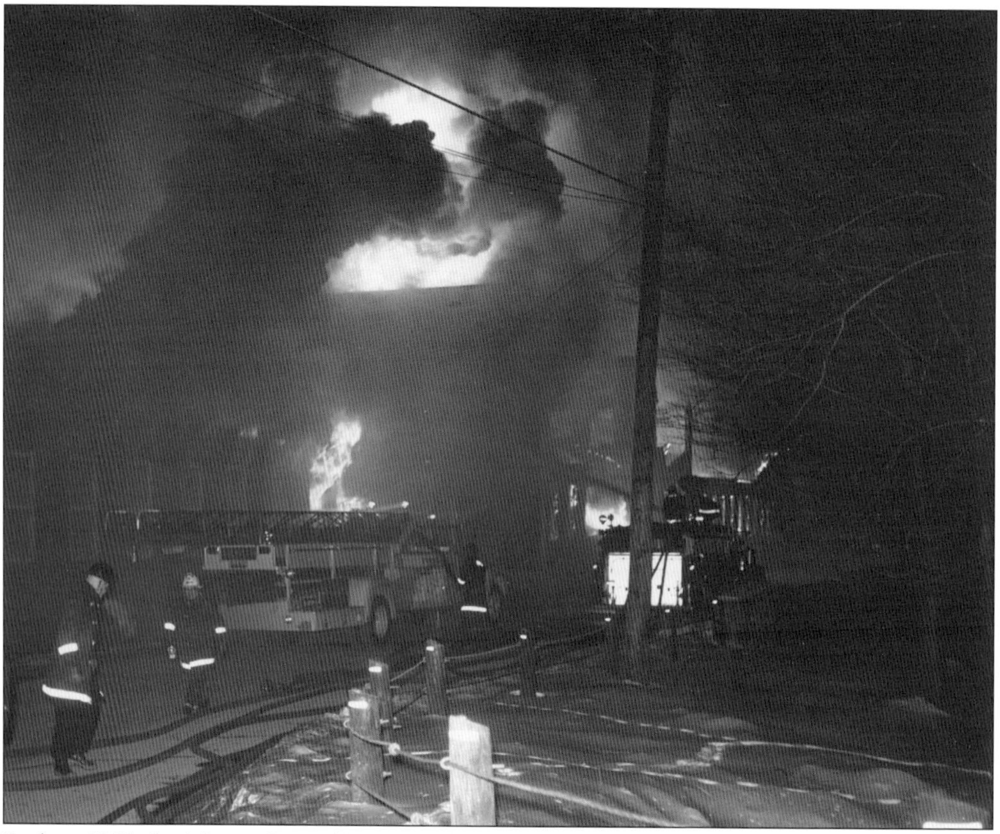

Built in 1835, the Mount Carmel Brass Works factory was located on Broadway at Ives Street. By the 1880s, it was W. Woodruff and Sons, manufacturer of some of the finest carriage hardware in the United States. By the mid-1960s, the building was occupied by the George T. Fleming Company, a manufacturer of tableware. At 3:45 a.m. on January 12, 1968, a truck driver miles away saw the glow in the sky and investigated. Arriving at the flaming building, he was the only one there. The fire had gone unnoticed. He raced to the nearby Mount Carmel fire station and turned in the alarm. For hours, dozens of firefighters fought extreme cold and frozen hydrants, saving part of the historic landmark. Firefighters successfully protected an adjacent house, literally a few feet away from the south end of the building. (Both, John Mongillo Jr.)

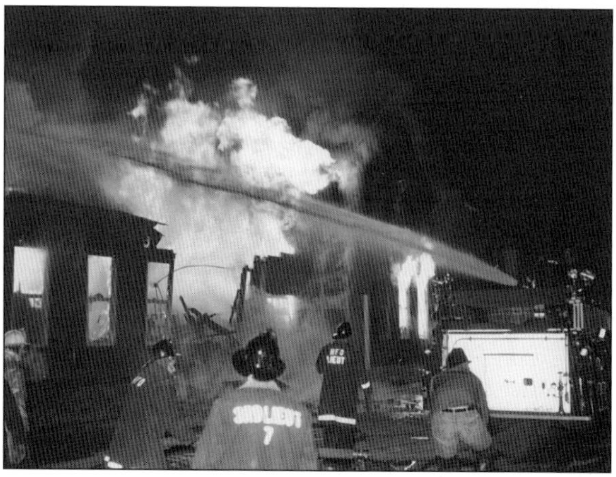

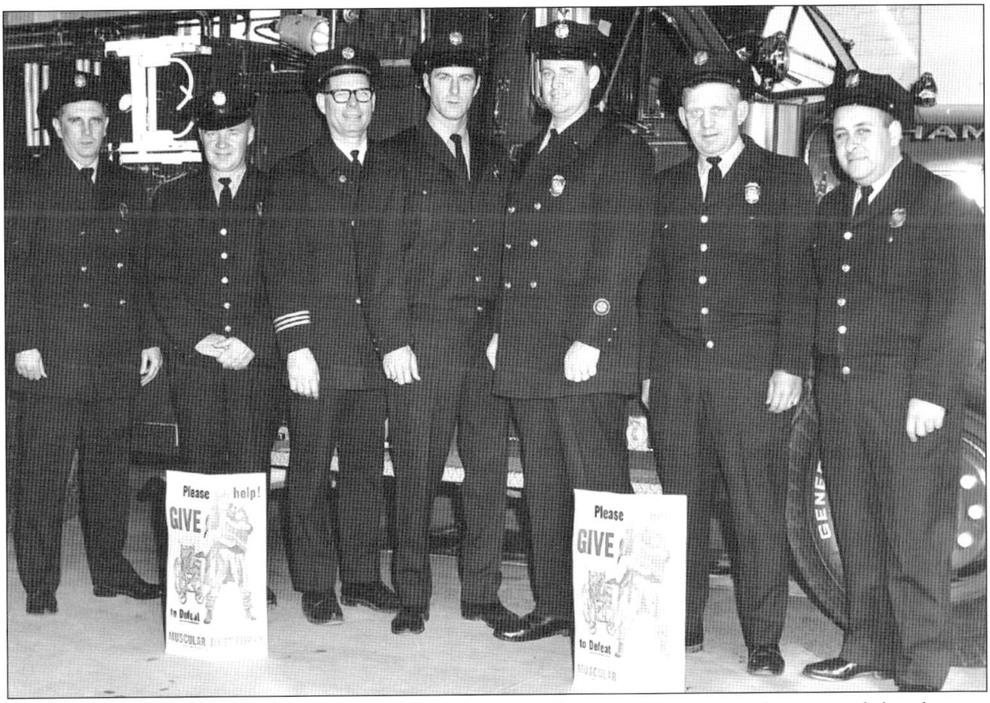

The Muscular Dystrophy Association (MDA) annual drive to raise money for research has been a favorite firefighter cause for decades. When over $4,300 was collected for the MDA's 1968 Mothers' March, members of the department's fundraising committee were, from left to right, Firefighters Raymond Vreeland and Milner Benham, Deputy Chief George Reutenauer, and Firefighters Paul Reutenauer, Gilbert Spencer, Robert Williams, and Charles Esposito. (HFRA.)

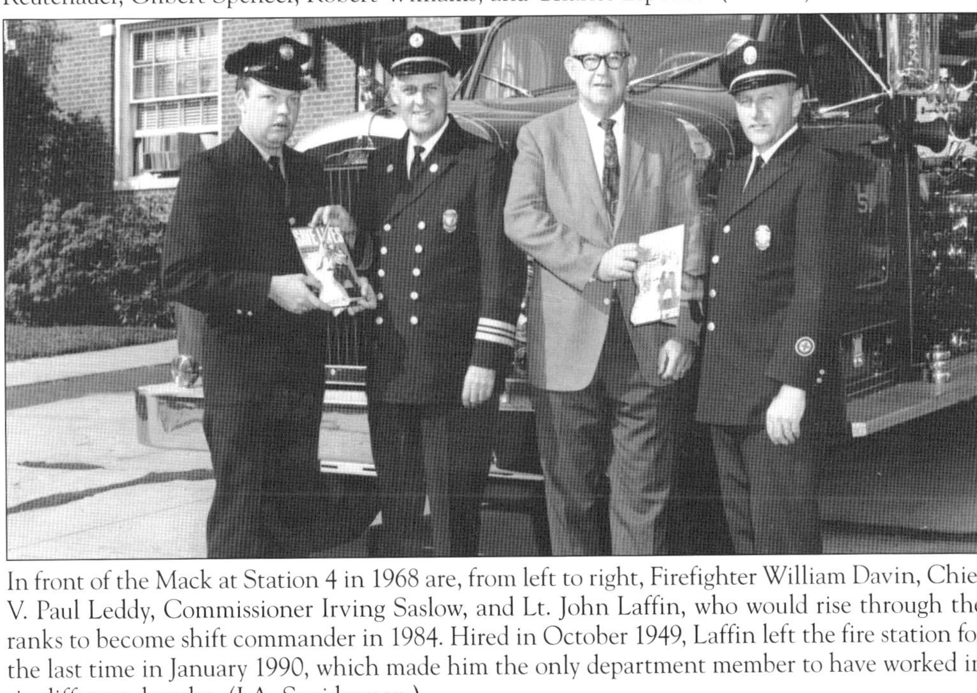
In front of the Mack at Station 4 in 1968 are, from left to right, Firefighter William Davin, Chief V. Paul Leddy, Commissioner Irving Saslow, and Lt. John Laffin, who would rise through the ranks to become shift commander in 1984. Hired in October 1949, Laffin left the fire station for the last time in January 1990, which made him the only department member to have worked in six different decades. (I.A. Sneiderman.)

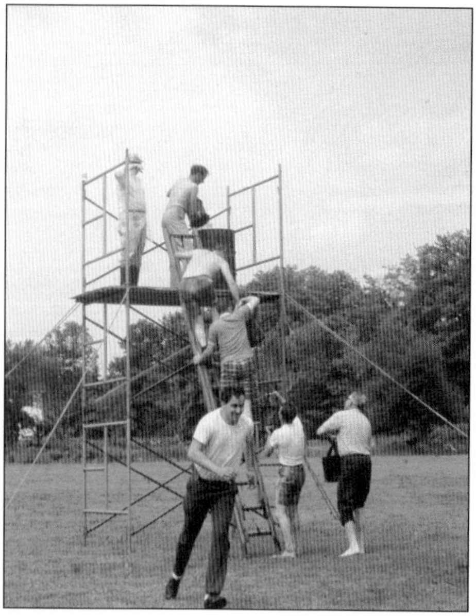

The annual Firemen's Field Day, held on the second Sunday of June during the 1960s, was a fun time for everyone. The four volunteer fire companies and the paid department each had a team in the competition. Trophies were awarded in hose laying, "midnight alarm," tug-of-war, bucket brigade, and ladder climbing. In the 1969 bucket brigade competition (at left), Richard Lostritto filled the 55-gallon drum from water bags grabbed by his teammates on the ladder, while Edward Doiron, foreground, and other teammates delivered full bags of water from a nearby trough. The team that filled the drum in the shortest time won. In the 1968 ladder competition (below), Mt. Carmel volunteers Ward Clemons Sr. and Ward Clemons Jr. footed the ground ladder while Richard Pleines, top, and David Johnson positioned the roof ladder for Pleines to climb to the ridge beam. (Left, Patricia Doiron; below, G. Donald Steele)

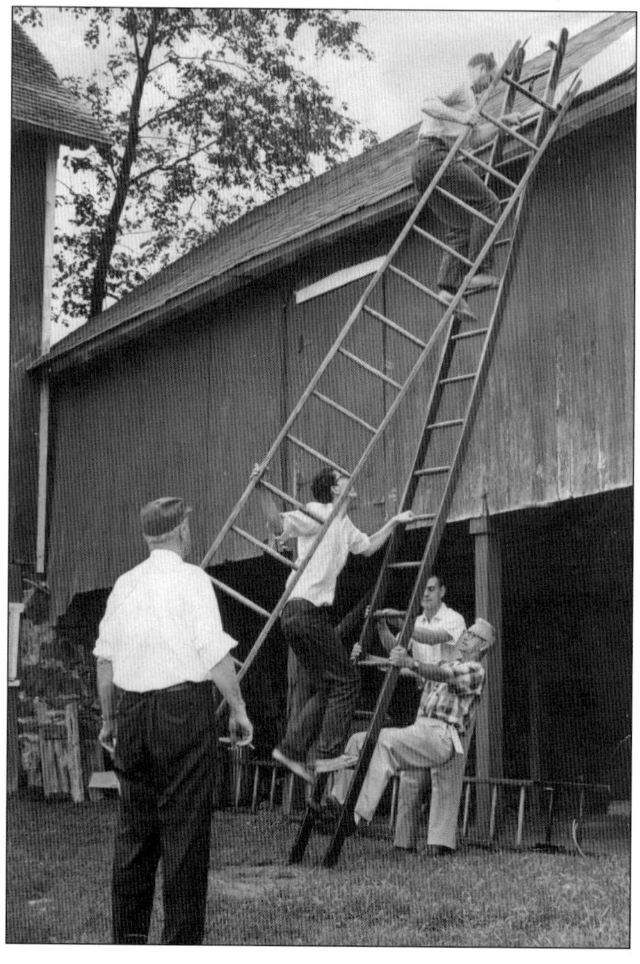

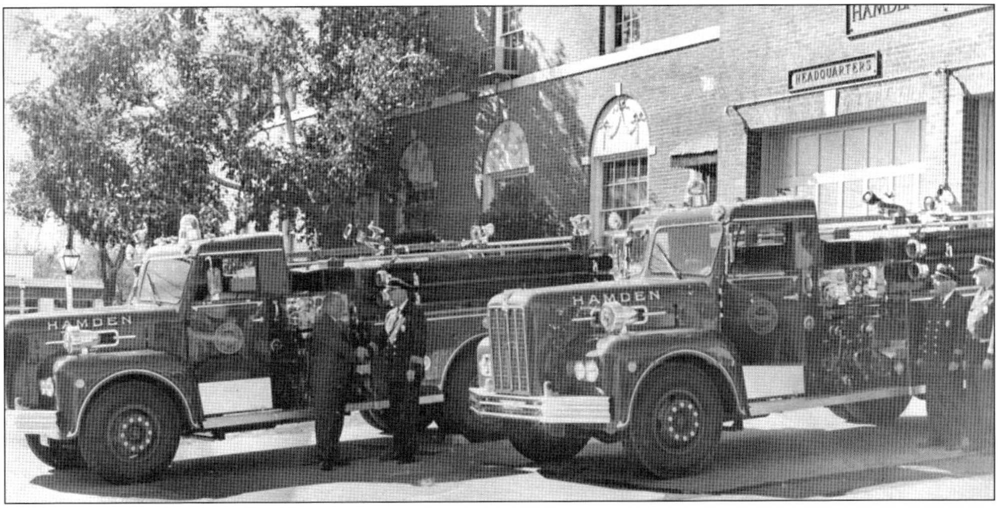

In October 1968, Hamden mayor William Adams (left) congratulated Chief V. Paul Leddy when the department accepted delivery of two identical Maxim 1,000-gpm S-Model pumpers that went in service as Engine 3 and Engine 4. On the right side of the image are Deputy Chief Training Officer Daniel Hume (left) and Supt. of Alarms and Apparatus Clement Wetmore. (John Mongillo Jr.)

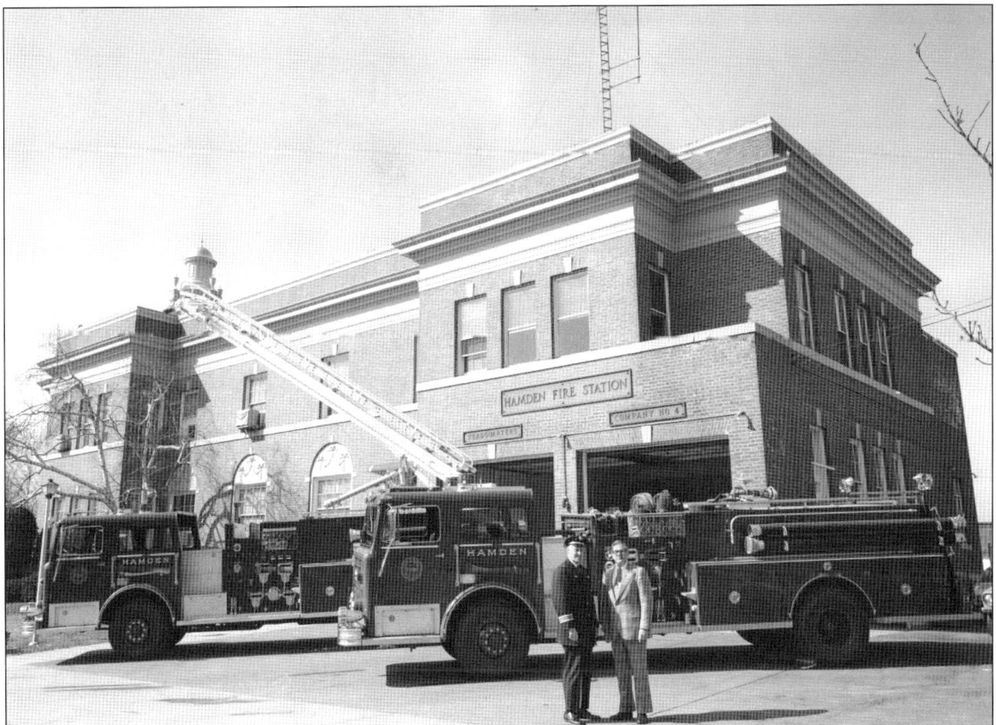

Another two pumpers were delivered in early 1974. Chief Leddy (left) and Hamden mayor Lucien A. DiMeo pose with new Engines 3 and 4, both Maxim 1,000-gpm pumpers. Engine 4 (foreground) was also equipped with a foam delivery system, turret nozzle, and ground sweeps. Engine 3, a TeleSqurt, was equipped with a pre-connected ladder pipe. The 1968 S-Models went to Station 2 as Engines 1 and 2. (John Mongillo Jr.)

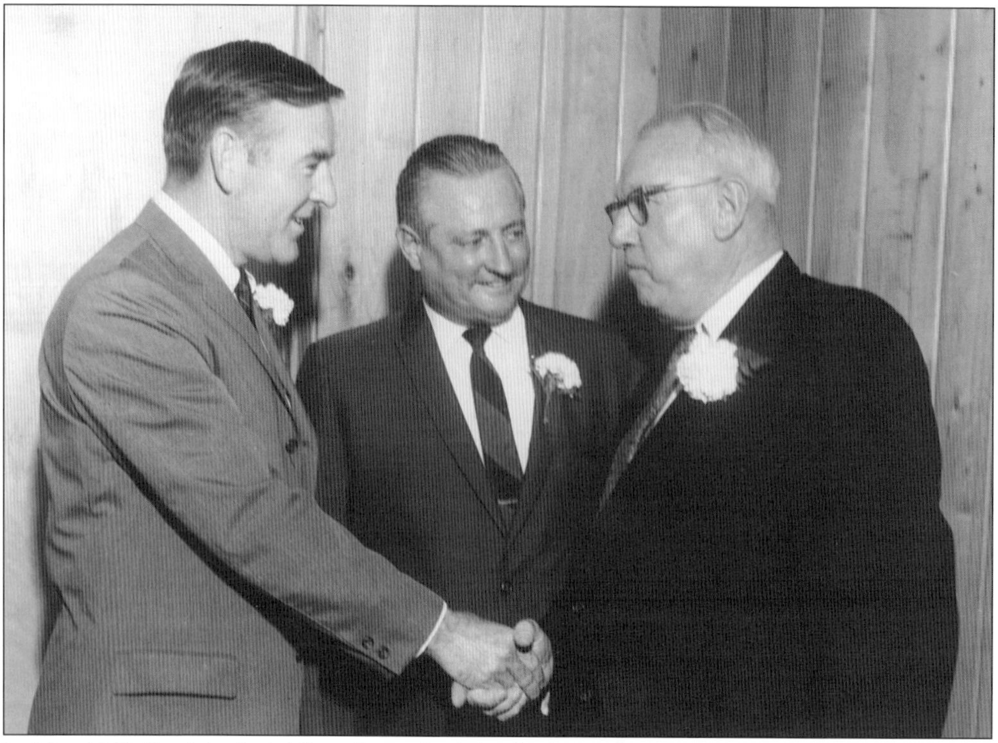

In May 1968, Fire Marshal Albert Purce (right), who began his career at Centerville in 1925 and was the first paid driver at Mount Carmel in 1926, retired after 43 years. While Mayor William Adams looks on, Purce shakes hands with his successor, 21-year veteran Capt. Robert "Bubby" O'Donnell, who would go on to serve as Hamden's second full-time fire marshal for nearly 20 years. (I.A. Sneiderman.)

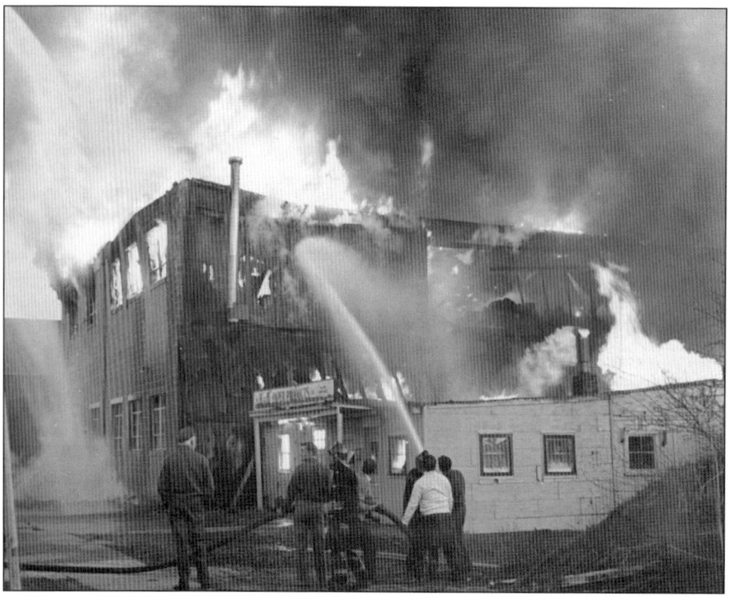

On Sunday, April 20, 1969, children playing with matches started this spectacular multi-alarm, $500,000 blaze that destroyed the H&D Paper Products Company on Webb Street, located right behind Merritt Street Station 6. Personnel from four Hamden engine companies and one from North Haven's West Ridge station fought the fire with the Hamden's ladder company and both rescues. (John Mongillo Jr.)

The ruins of the H&D Paper Products Company are still visible behind Merritt Street Station 6 in this photograph taken soon after the fire. The fire station, owned by the Merritt Street Volunteer Fire Co. No. 6 since it was built in 1922, closed in September 1970, when "new" Station 3 opened just over a mile away at Ridge Road and Hartford Turnpike. (Edward Doiron Sr.)

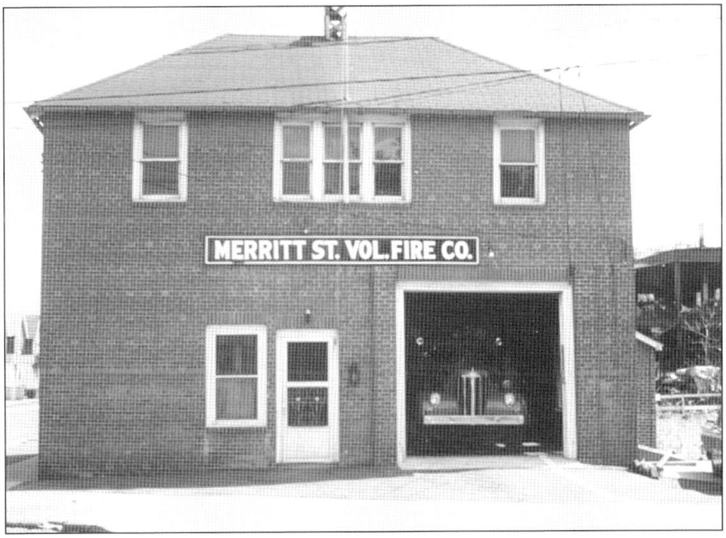

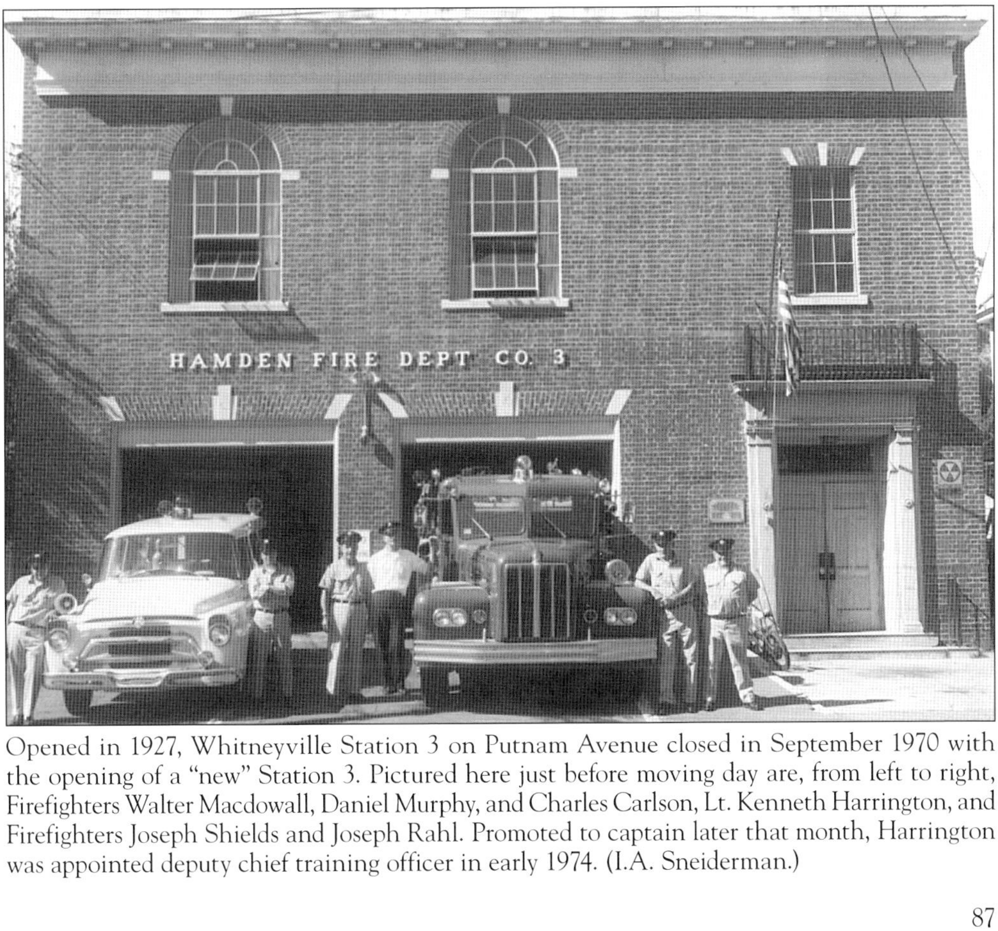

Opened in 1927, Whitneyville Station 3 on Putnam Avenue closed in September 1970 with the opening of a "new" Station 3. Pictured here just before moving day are, from left to right, Firefighters Walter Macdowall, Daniel Murphy, and Charles Carlson, Lt. Kenneth Harrington, and Firefighters Joseph Shields and Joseph Rahl. Promoted to captain later that month, Harrington was appointed deputy chief training officer in early 1974. (I.A. Sneiderman.)

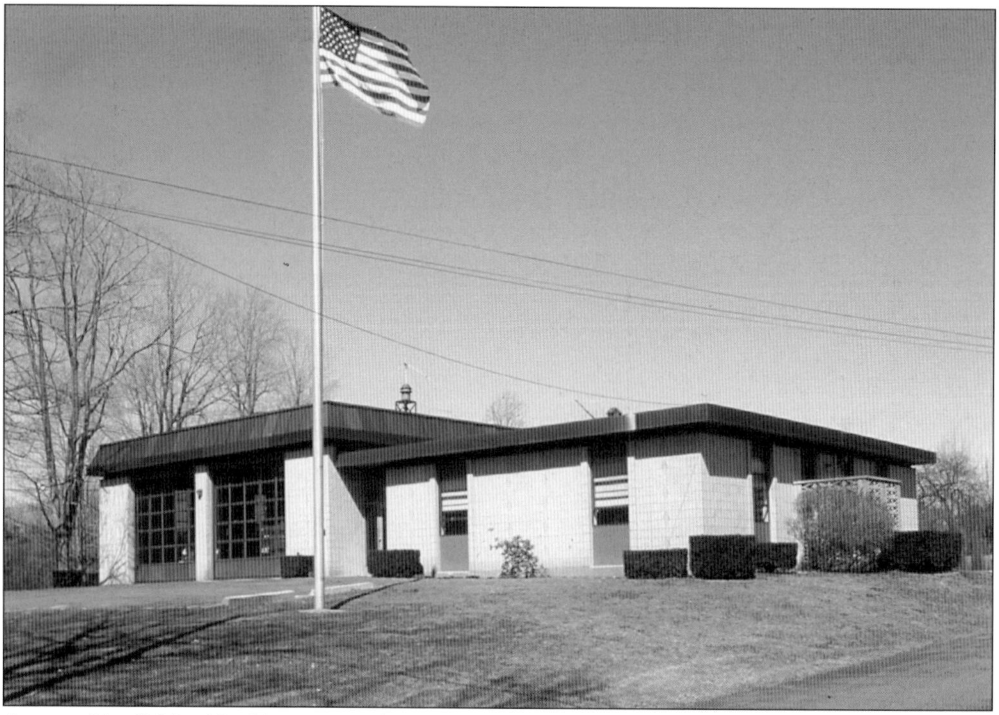

Among Chief V. Paul Leddy's many achievements during his tenure was overseeing the construction of two new fire stations. Station 9 (above), at the corner of Johnson and Still Hill Roads in West Woods, was dedicated in December 1968 for Engine 9, a new engine company manned by career personnel. The West Woods volunteers remained active, their quarters now farther back on the property. With the closing of Whitneyville Station 3 and Merritt Street Station 6, a new Station 3 (below), opened at the corner of Hartford Turnpike and Ridge Road in September 1970. Assigned to the new station were Engine 3, Engine 6, Rescue 1, Car 30 (shift commander), and Ladder 2, a brand-new Maxim 100-foot aerial ladder truck. New offices were provided for the shift commanders and station officers. (Above, John Tramontano; below, Calvin Stoner.)

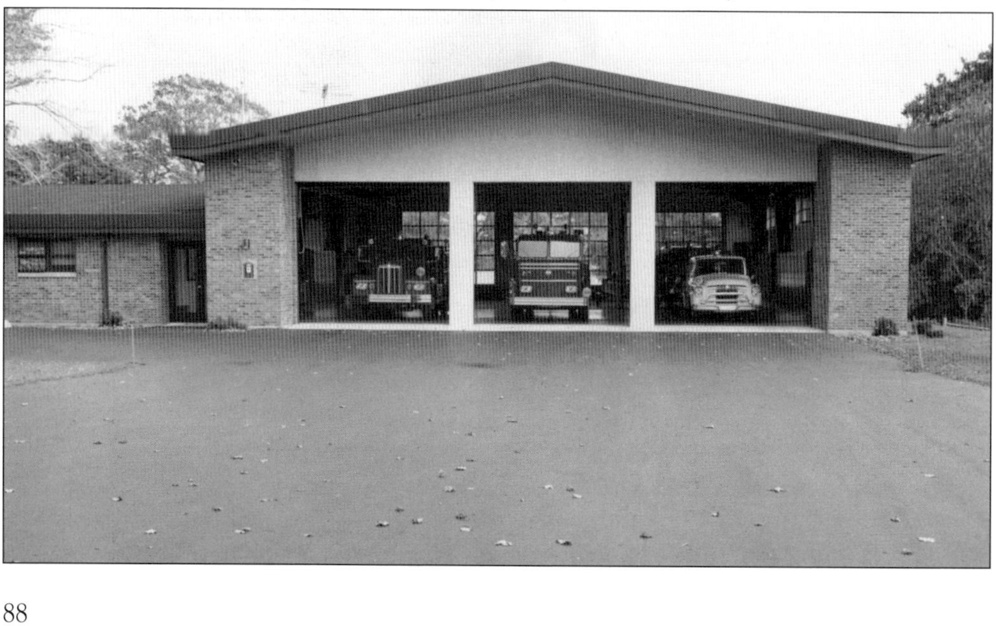

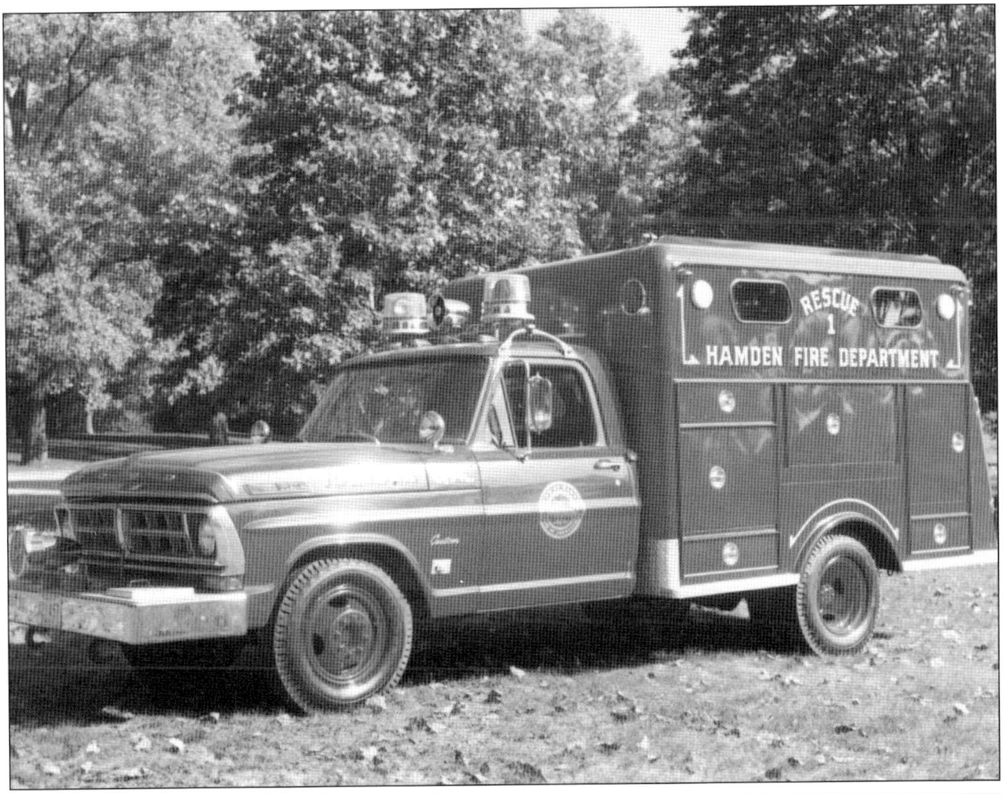

With the certification of the department's first 17 emergency medical technicians (EMTs), the two aging International-Harvester "rescues" were retired, and this new modular unit on a Ford chassis was placed in service as Rescue 1 in November 1971. It was assigned to Centerville Station 4. (HFRA.)

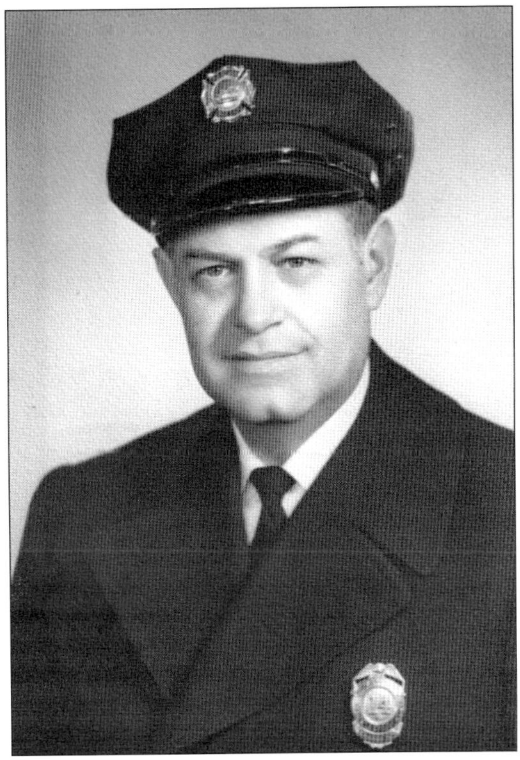

On Christmas Eve morning 1971, Firefighter Alfred "Al" Ramelli was stricken in the fire station and died on New Year's Day 1972. He was survived by his wife Helyn, son Raymond (also a Hamden firefighter), daughter-in-law Tina, and grandsons Raymond Jr., Roger, and Al. At the department's first awards ceremony in 1985, Firefighter Raymond Ramelli accepted the department's Medal of Honor, awarded posthumously to his father. (Ramelli family.)

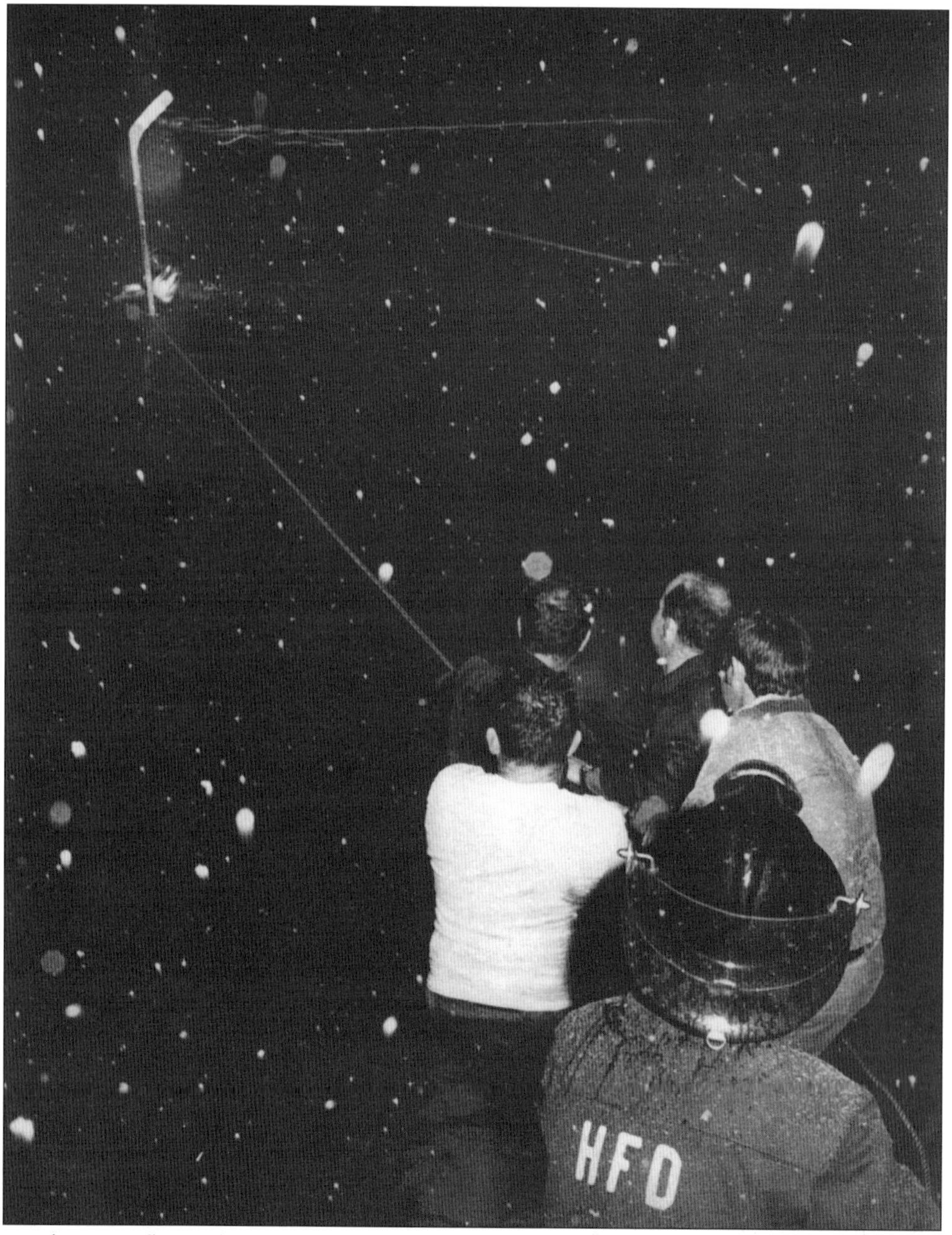

As early winter flurries began, young Brad Crerar and a friend were playing hockey after school on Peter's Pond off Mix Avenue. After they both fell through the ice, Brad's friend got to safety and notified the fire department. This remarkable photograph of Hamden Fire Department personnel rescuing the young man was taken by legendary news photographer John Mongillo Jr. Initially published in the *New Haven Register*, the photograph was picked up by the Associated Press and featured in the February 4, 1972, issue of *Life* magazine. Fire personnel are, clockwise from the top, Firefighters Gerald Wolf, Edward Charbonneau, Walter Macdowall, and Robert Viglione (in turnout gear). The fellow in the white T-shirt is Volunteer Firefighter Wesley DeGroot of Mix District Co. 7. (Photograph by John Mongillo Jr. reprinted here with the permission of the *New Haven Register*.)

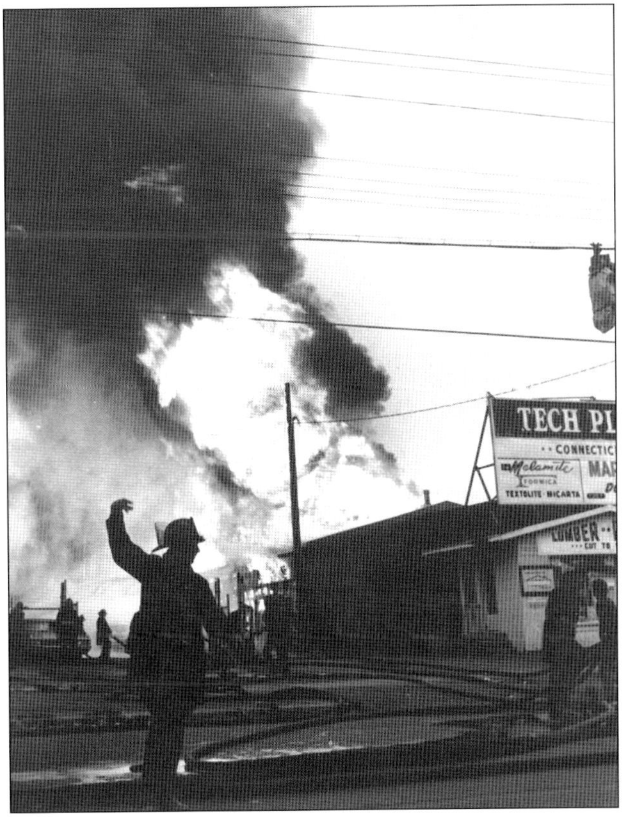

In the late afternoon of October 17, 1972, the Tech Plywood fire at 1047 Dixwell Avenue went to two alarms. In this photograph, Firefighter Douglas Yocher of Engine 3 directs Ladder 2 into position in front of the building. Firefighters saved the company's offices in front, but fire destroyed the entire stock of lumber and several automobiles parked at Aamco Transmission next door. (John Mongillo Jr.)

On February 24, 1973, the cab of this tanker truck caught fire while the driver was delivering oil. The fire extended to the gasoline tank, which then exploded and ignited the cargo of home heating oil. The home at 86 Glenbrook Avenue was heavily damaged, but firefighters contained and extinguished the fire quickly, preventing extension to the surrounding houses. (HFRA.)

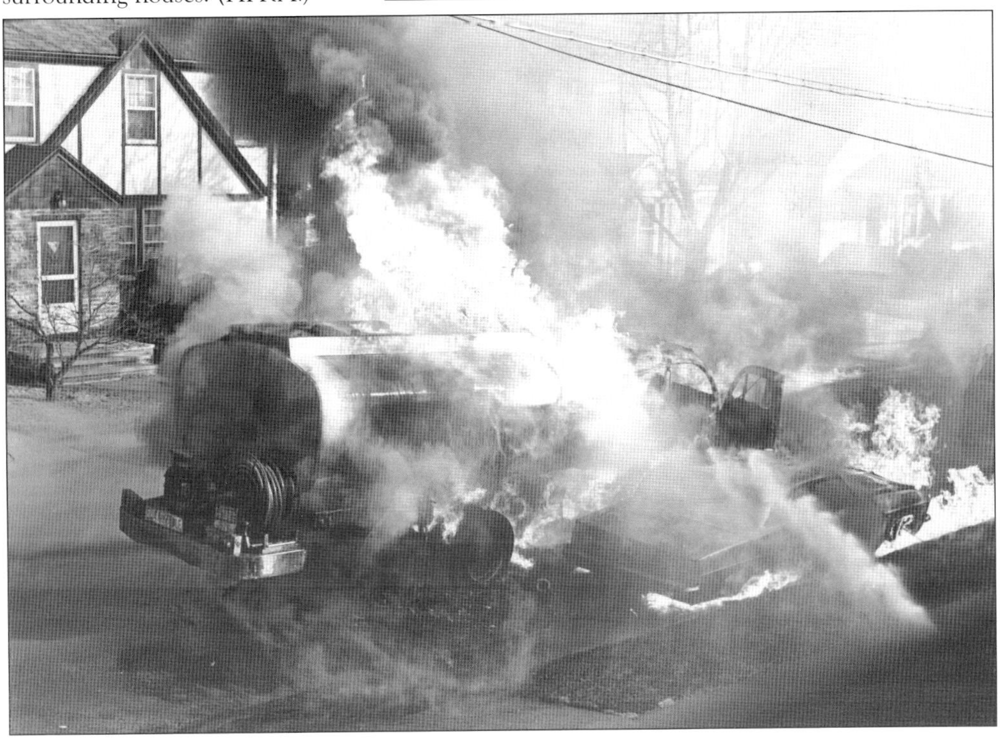

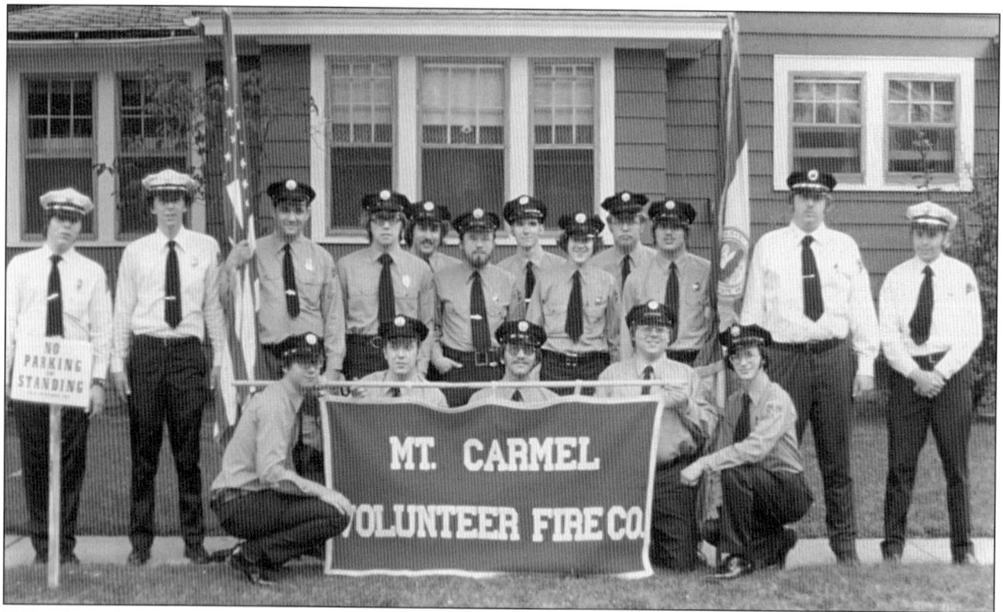

Whether they are career or volunteer, firefighters always enjoy summer parades and musters. In June 1975, members of the Mt. Carmel Volunteer Fire Co. No. 5 attended the Southern New York Volunteer Firemen's Convention Parade at Lynnbrook, Long Island. At least six members in this photograph either were, or would become, career firefighters. (Edward Doiron Sr.)

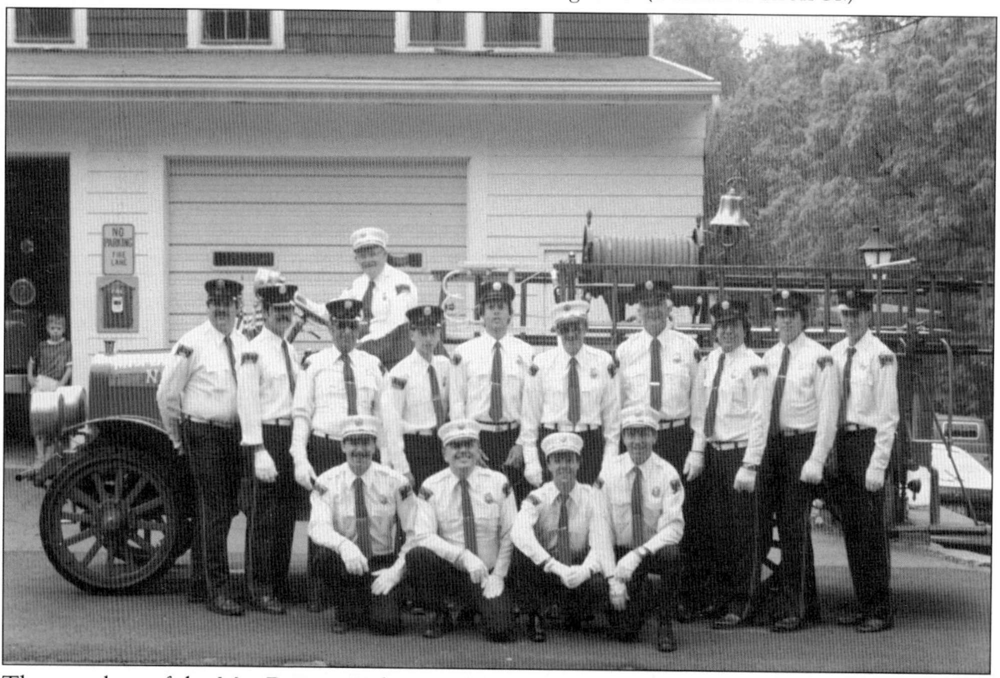

The members of the Mix District Volunteer Fire Co. No. 7 are preparing for a 1980s parade. At the wheel of the company's 1918 Brockway chemical truck is charter member and Supt. of Alarms and Apparatus Clement Wetmore (retired). His extended family included many Co. 7 members, as well as career members, including his son Paul, who served as Hamden fire chief from 1993 to 1996. (Erwin Wetmore.)

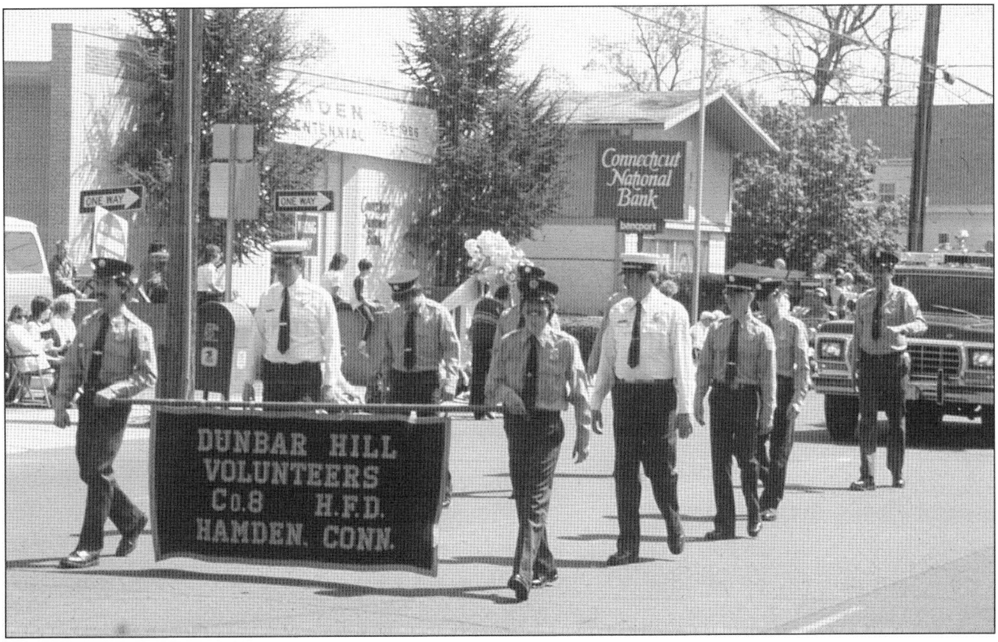

Members of the Dunbar Hill Volunteer Fire Co. No. 8, one of three Hamden volunteer companies still active in 2017, were participants in the May 11, 1986, Hamden Bicentennial Parade along Whitney Avenue. Capt. Timothy Sullivan of Co. 8, marching immediately behind the company's banner, would serve as Hamden's fire chief from 1996 until his 2002 retirement. (Edward Doiron Sr.)

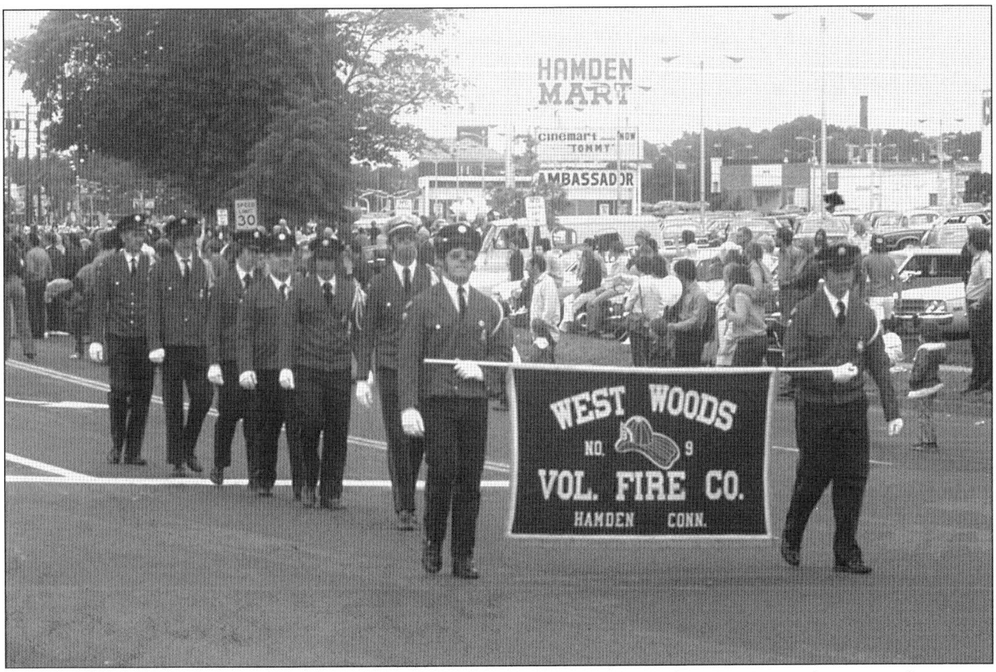

The West Woods Volunteer Fire Co. No. 9 organized in 1956 and was extremely active for many years, even after the opening of the career fire station at West Woods in 1968. The West Woods volunteers are pictured here marching in Hamden's 1975 Memorial Day Parade along Dixwell Avenue near the Hamden Mart. (Edward Doiron Sr.)

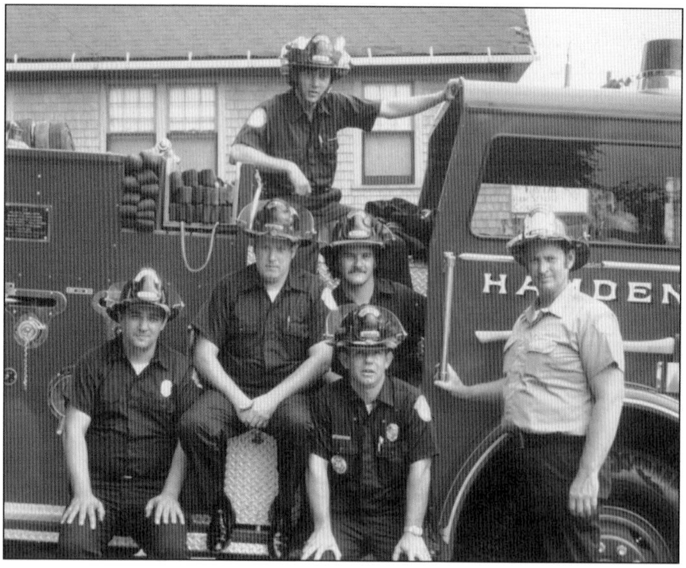

In July 1974, the crew of Station 4, Platoon 2 pose with Hamden's new Maxim 1,000 pumper, delivered a few months earlier. Up top is Firefighter Steven Hitchcock. Seated are, from left to right, Firefighters Edward Doiron Sr., Robert Kenney, Robert Viglione, and Edward Charbonneau (in front of Viglione). Standing is Lt. Gilbert Spencer, who would advance to captain and battalion chief before retiring in 1992. (Edward Doiron Sr.)

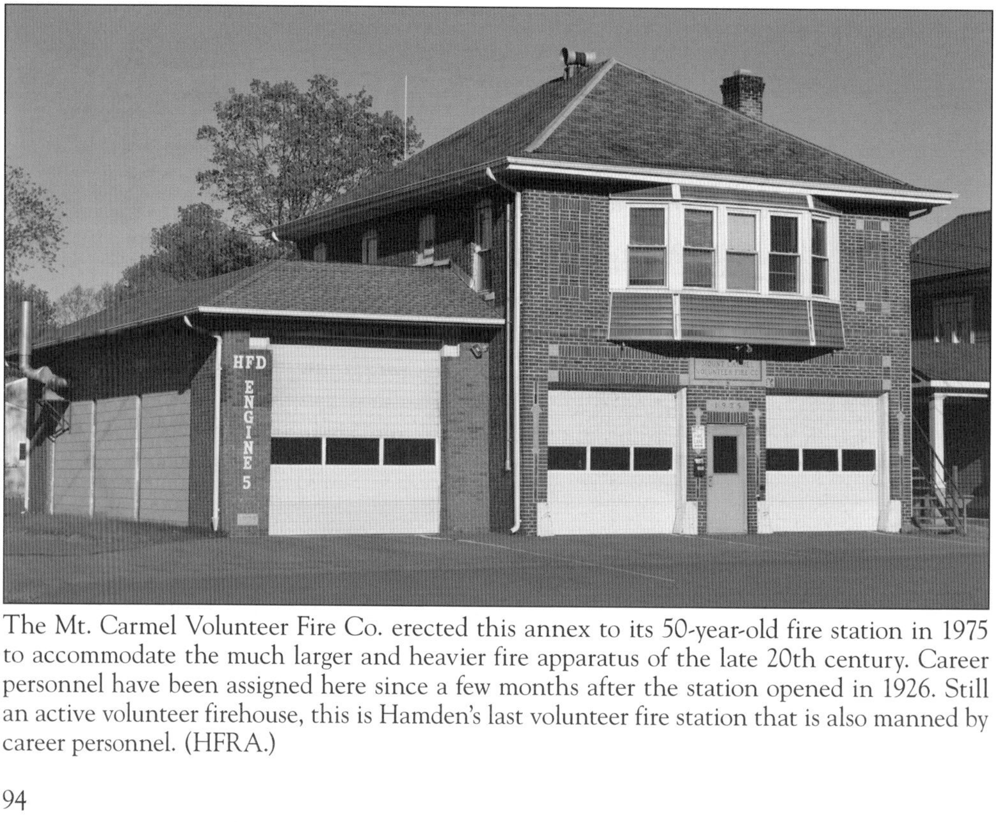

The Mt. Carmel Volunteer Fire Co. erected this annex to its 50-year-old fire station in 1975 to accommodate the much larger and heavier fire apparatus of the late 20th century. Career personnel have been assigned here since a few months after the station opened in 1926. Still an active volunteer firehouse, this is Hamden's last volunteer fire station that is also manned by career personnel. (HFRA.)

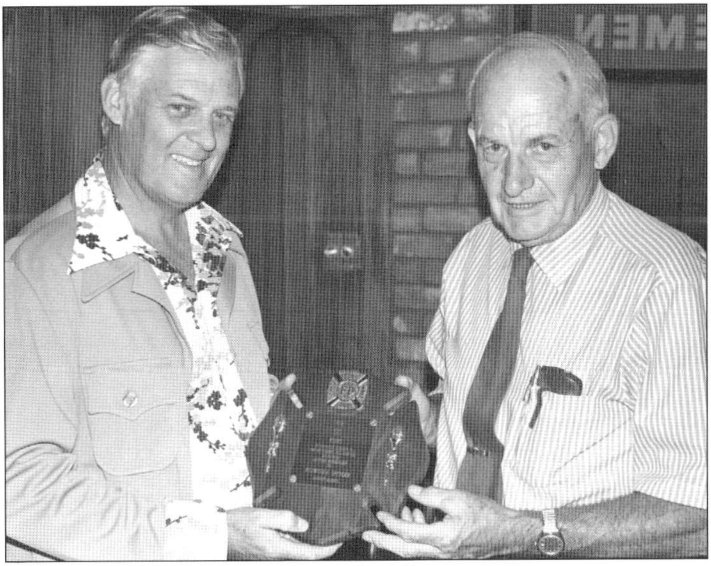

In August 1975, Raymond K. Spencer (right) was honored at New Haven's Firehouse Restaurant for his 50 years of service with the Mt. Carmel volunteers. Mt. Carmel lifetime member Chief V. Paul Leddy presented Spencer with this plaque honoring his service, which continued for more than 10 years. On September 4, 2011, the fire company named the 1975 Station 5 annex in memory of Spencer. (Jack Harriett.)

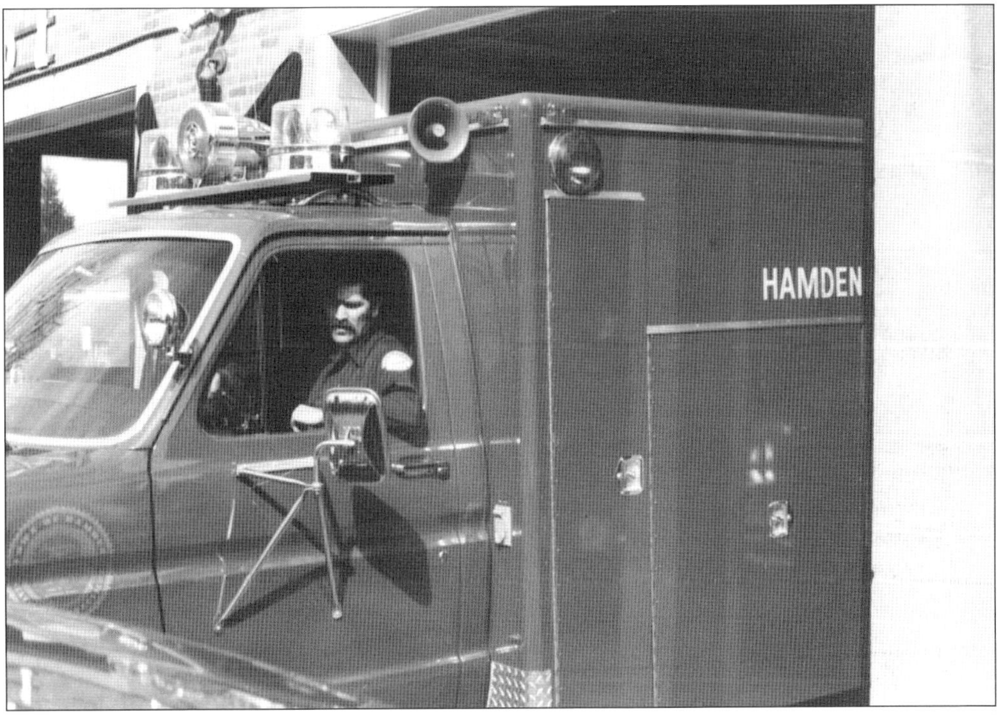

After Truck 1 was reassigned to the new Mt. Carmel annex on April 9, 1976, Firefighter Raymond Reilly backed new Rescue 2 into Humphrey Station 2. The inauguration of this new rescue unit, donated by Hamden's UNITAS Club, coincided with the certification of the department's first eight paramedics: Lts. Walter Macdowall and John Tramontano and Firefighters Milner Benham, John Calamo, Edward Charbonneau, Charles Esposito, Howard Hurlburt Jr., and Robert Westervelt. (HFRA.)

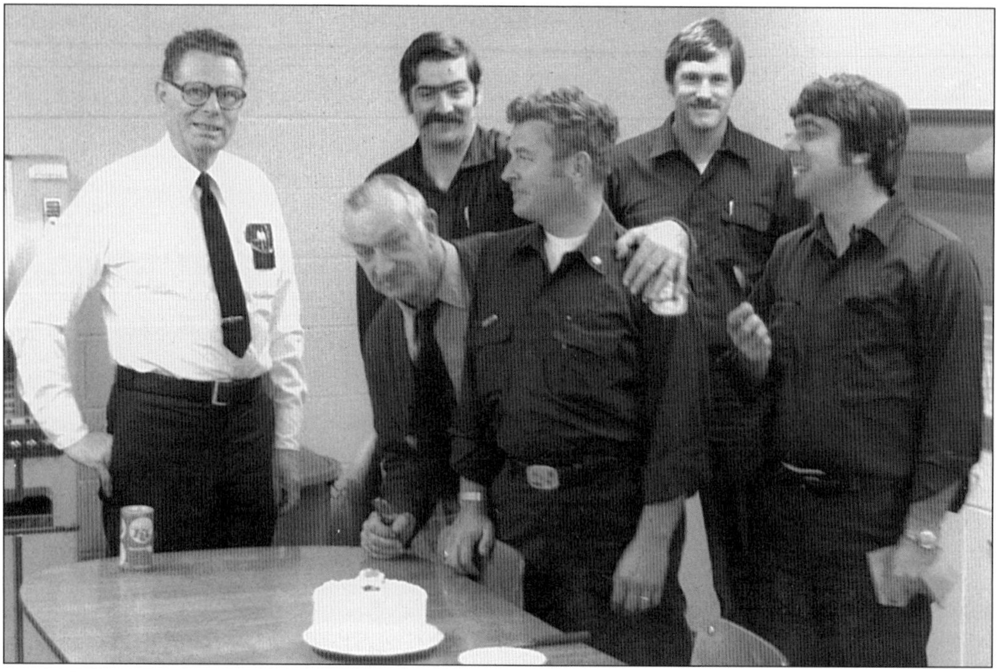

Firefighter Guy White left the department in 1978 to take a position with an insurance company in Georgia. A firefighter's crew is like a second family, so White's buddies at Station 3 held an impromptu farewell party on his last day. From left to right are Deputy Chief George Reutenauer, Lt. Francis Eitler, and Firefighters Raymond Reilly, White, Jeffrey Stoehr, and Daniel Murphy. (David Johnson.)

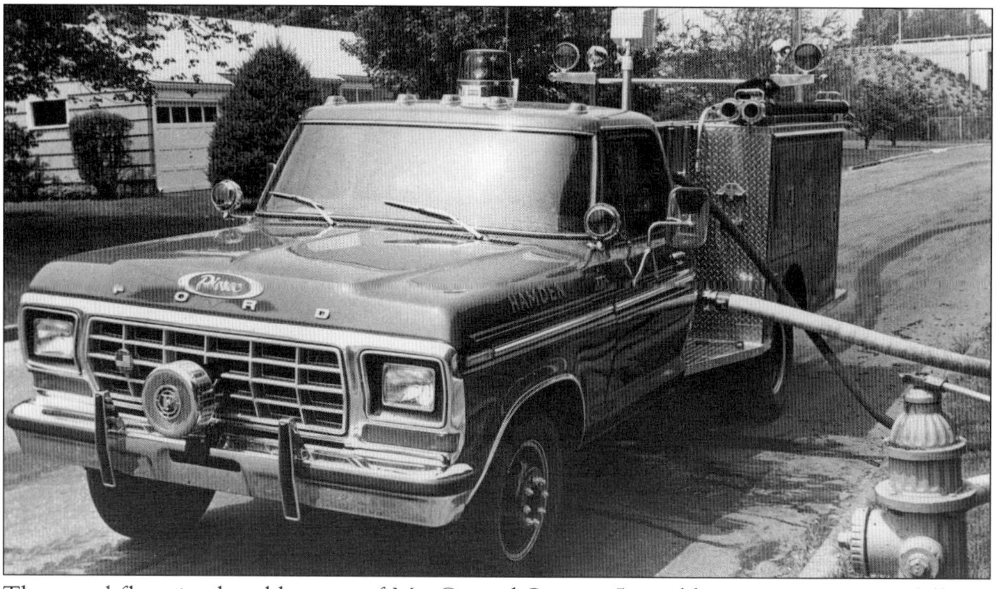

The wood floor in the older part of Mt. Carmel Station 5 would not support a new full-size pumper, so this new Pierce/Ford 450-gpm mini-pumper was delivered to Mt. Carmel in 1978. When the ladder truck moved from Station 5's annex to Station 9 in 1985, a 1968 Maxim took its place in the annex and the mini-pumper went to volunteer Co. 8. (Edward Doiron Sr., *Alarm Room News*.)

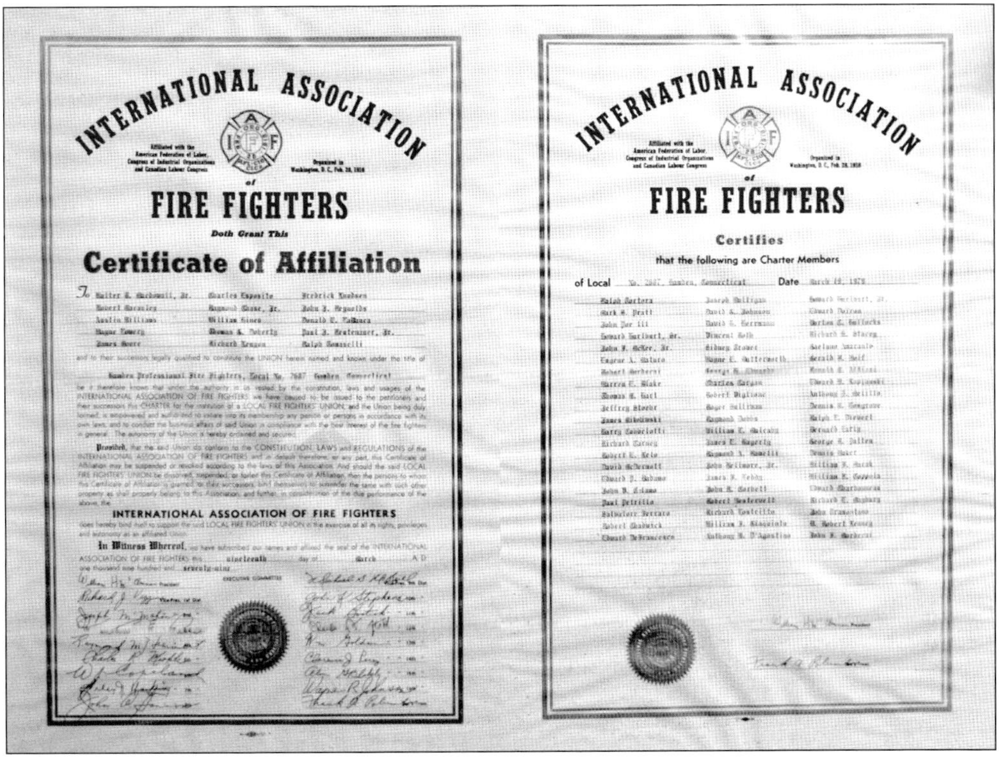

By the late 1970s, inflation was skyrocketing nationwide and the town was stalling in contract negotiations with the Hamden Paid Firemen's Sick Benefit Association, the firefighters' bargaining unit since the 1960s. After the current labor contract expired in 1978, nonmanagement department personnel formed the Hamden Firefighters' Association (HFA). Finally, after working without a raise for nearly two years, and with the town still stalling on negotiations, the HFA members voted overwhelmingly to join the International Association of Fire Fighters (IAFF). In March 1979, the Hamden Professional Firefighters Association Local 2687 became the rank and file's official bargaining unit. Firefighter John Corbett was elected the local's first president. A new contract was signed soon thereafter. Corbett served as Local 2687's president for 13 years. He was succeeded in 1993 by Firefighter Robert Anthony. After his retirement, Anthony ran for mayor of Hamden in 2013 and 2015. (Both, HFRA.)

Hamden Professional Firefighters Association
Local 2687, I.A.F.F.

Association Presidents

John Corbett
1979 - 1992

Robert Anthony
1993 - 2004

Gary Merwede
2005 - 2011

Kurt Vogt
2012 -

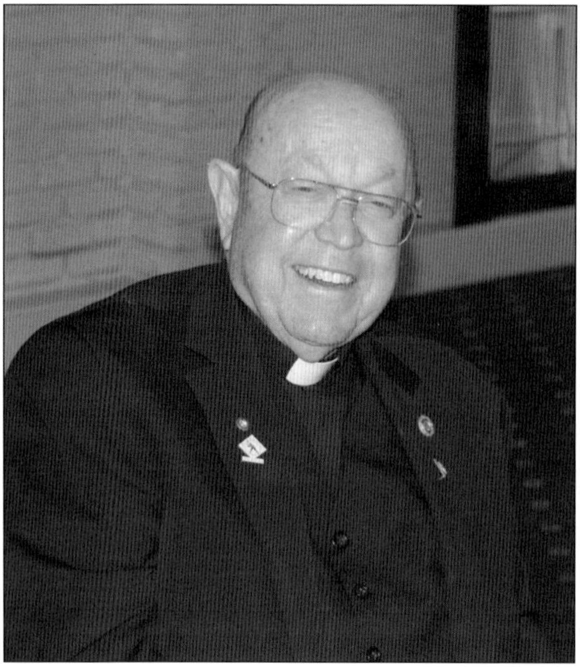

Rev. Owen Sanderson of Hamden's Christ Lutheran Church has been chaplain of the department since he was sworn in at the dedication of Hamden's new Station 3 in November 1970. Affectionately known as "the Rev," he is always found at active Hamden emergencies to help the troops and, when necessary, to console the afflicted. He is the only regular member of the HFRA who is not actually retired. (HFRA.)

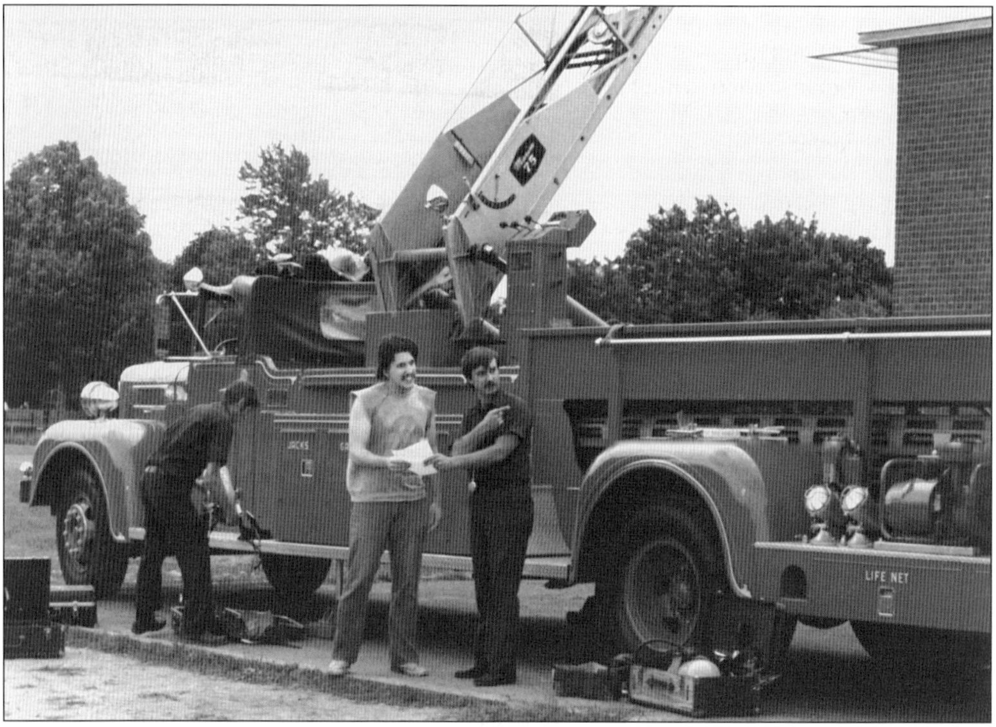

By the late 1970s, physical fitness testing for recruit firefighter candidates began to include job-related tasks. In this August 1980 photograph, a candidate has completed the climb of the 75-foot aerial ladder within the required time limit while wearing a Scott airpack. Firefighter Ernest Braun (left) is preparing for the next candidate, while Firefighter Gregory Shammett directs the candidate to the next test station. (Harold Wheatley.)

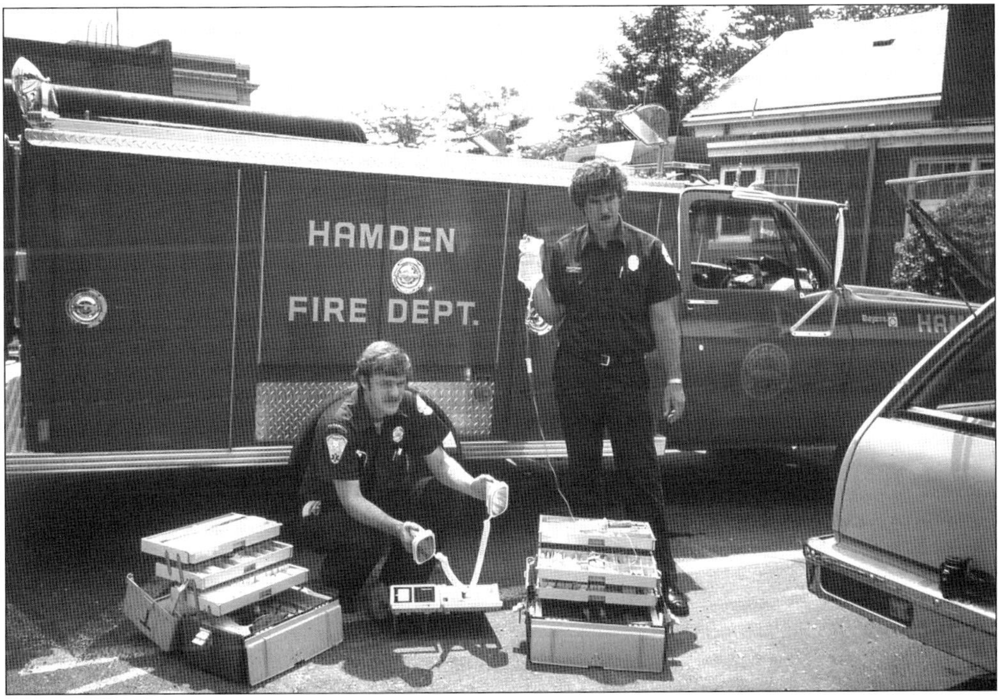

Firefighter-paramedics Howard Hurlburt Jr. (left) and Robert Surprise conducted a demonstration behind Central Communications in May 1982. Son of Firefighter Howard Hurlburt Sr., the younger Hurlburt was among the department's first eight paramedics in 1976. Firefighter Surprise rose through the ranks to become the department's second-in-command, succeeding Hurlburt's younger brother Clark, who served as the department's deputy chief from 2006 until his retirement in 2010. (John Tramontano.)

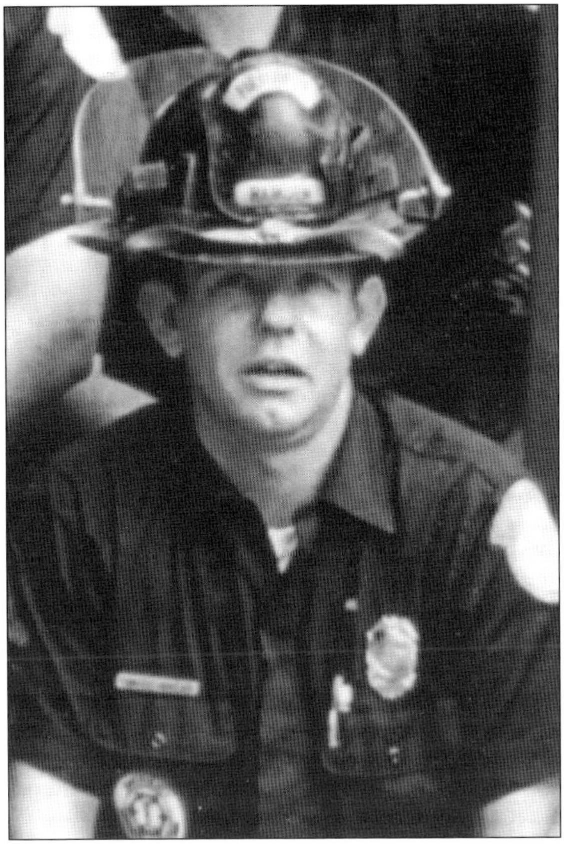

Firefighter Edward "Eddie" Charbonneau joined the department in 1968. Three years later, he was among the department's first EMTs and, in 1976, one of the department's first certified paramedics. On Saturday, December 4, 1982, Charbonneau died suddenly while on maneuvers with the US Army Reserve in Windsor Locks. Only 43 years old, he was survived by his wife, Joan, and two children. (Edward Doiron Sr.)

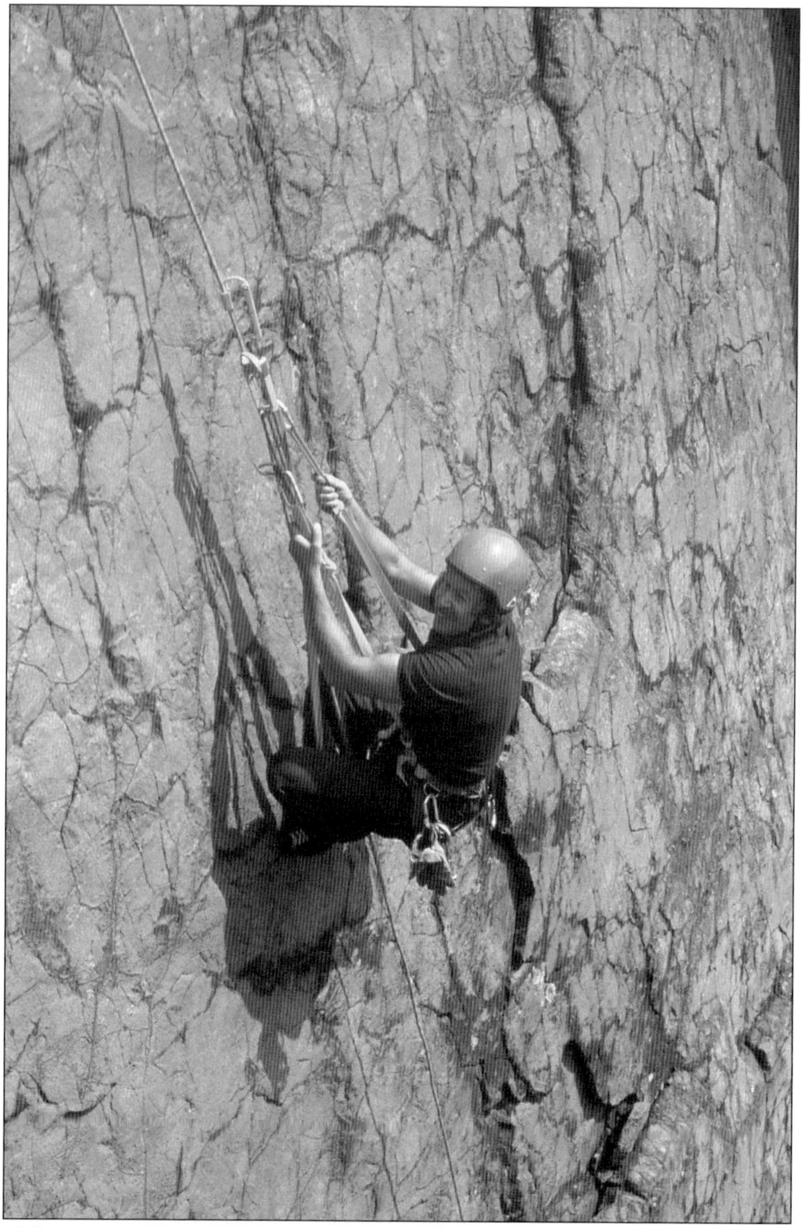

Since the 19th century, the mountain in northern Hamden known as Sleeping Giant has attracted many a would-be mountain climber, sometimes with tragic consequences. Inexperienced amateurs would attempt to climb the sheer cliffs of what was once an active quarry, eventually finding themselves unable to either ascend or descend, placing themselves and their rescuers in peril. John Tramontano and William Coppola, both experienced mountain climbers, recognized the need to have trained climbers to coordinate and accomplish such high-elevation rescues, not only on "the Giant" but also elsewhere. In 1976, Lieutenant Tramontano and Firefighter Coppola organized the department's Mountain Rescue Team. They trained fellow team members at Sleeping Giant and elsewhere. In over 40 years, the Mountain Rescue Team has saved many climbers in Hamden and area communities. This 1982 photograph shows Deputy Chief Tramontano rappelling from a very high elevation in Southington. (HFRA.)

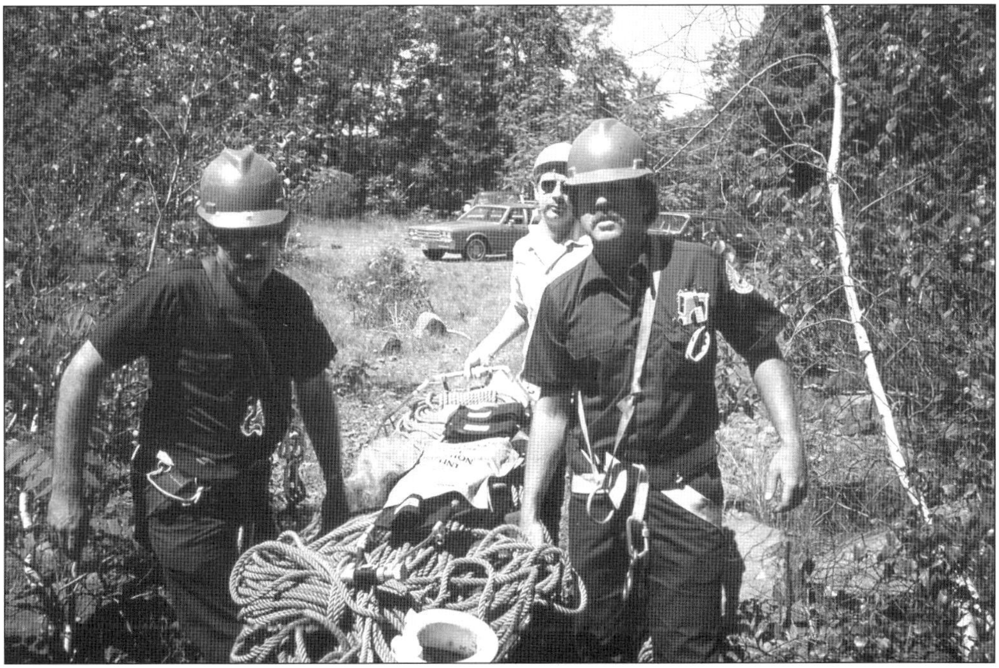

At the Sleeping Giant quarry, Mountain Rescue Team members Firefighters Anthony Melillo (left), Timothy O'Flynn, and Thomas Hart carry the team's gear that is stored on one of the department's rescue units. The quarry is where most of the Hamden rescues take place. Hapless climbers get halfway up and then realize they cannot proceed up or down. The team's services have been rendered in other area communities as well. (HFRA.)

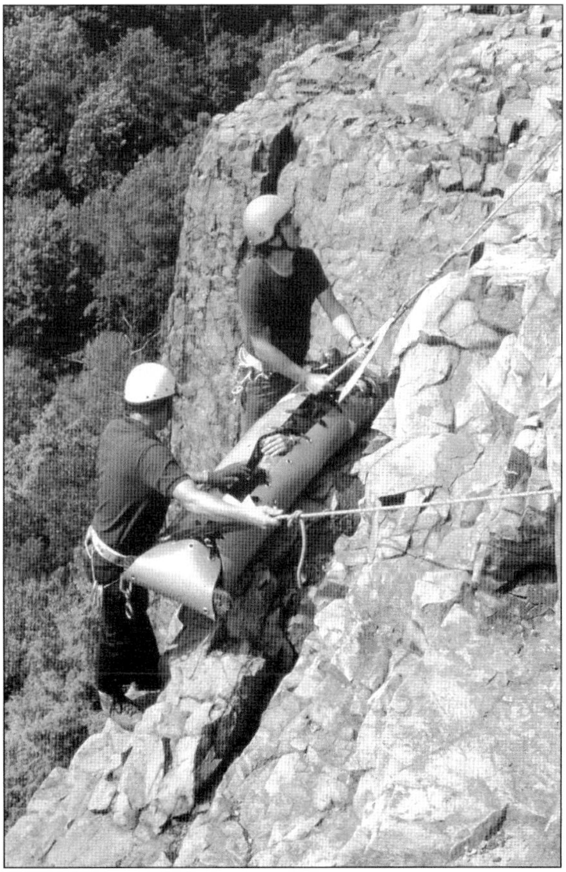

In this 1987 training exercise in Southington, the "victim," firefighter and future fire marshal Mark Barletta, has been stabilized and "packaged" in a Sked stretcher. Firefighter Fred "Chick" Manware (left) and Capt. William Coppola are preparing to lower him to safety over 100 feet below. (HFRA.)

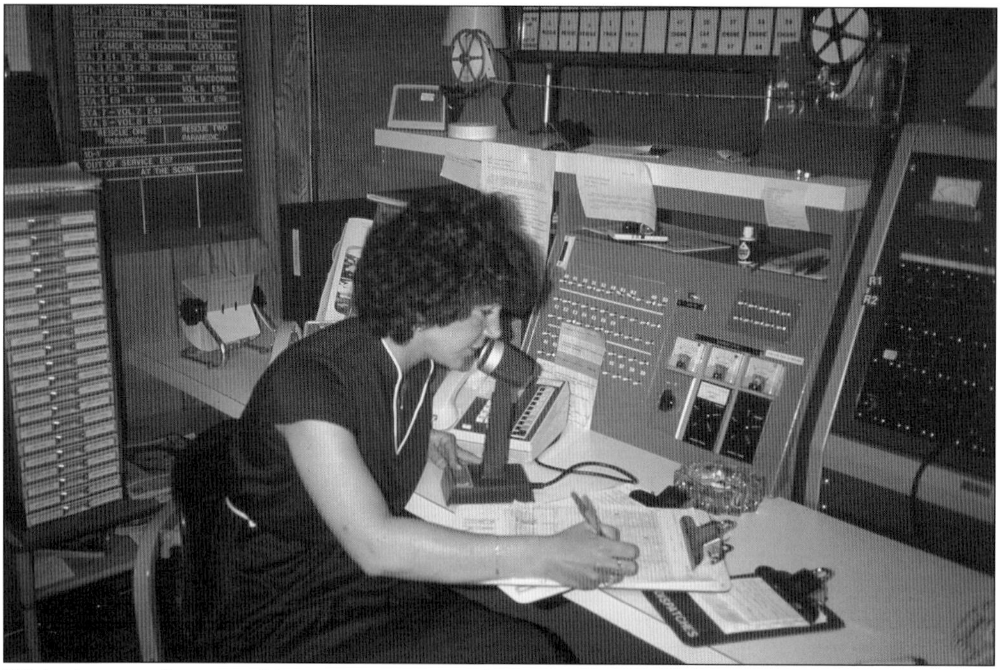

With the introduction of 911 service in Hamden on November 19, 1981, the dispatch facilities of Hamden's police and fire departments were combined into Central Communications. Former civilian police and fire dispatchers were cross-trained to handle emergency calls under the protocols of each department. Dispatcher Marge Gambardella is pictured here at the fire dispatch console in April 1982. (John Tramontano.)

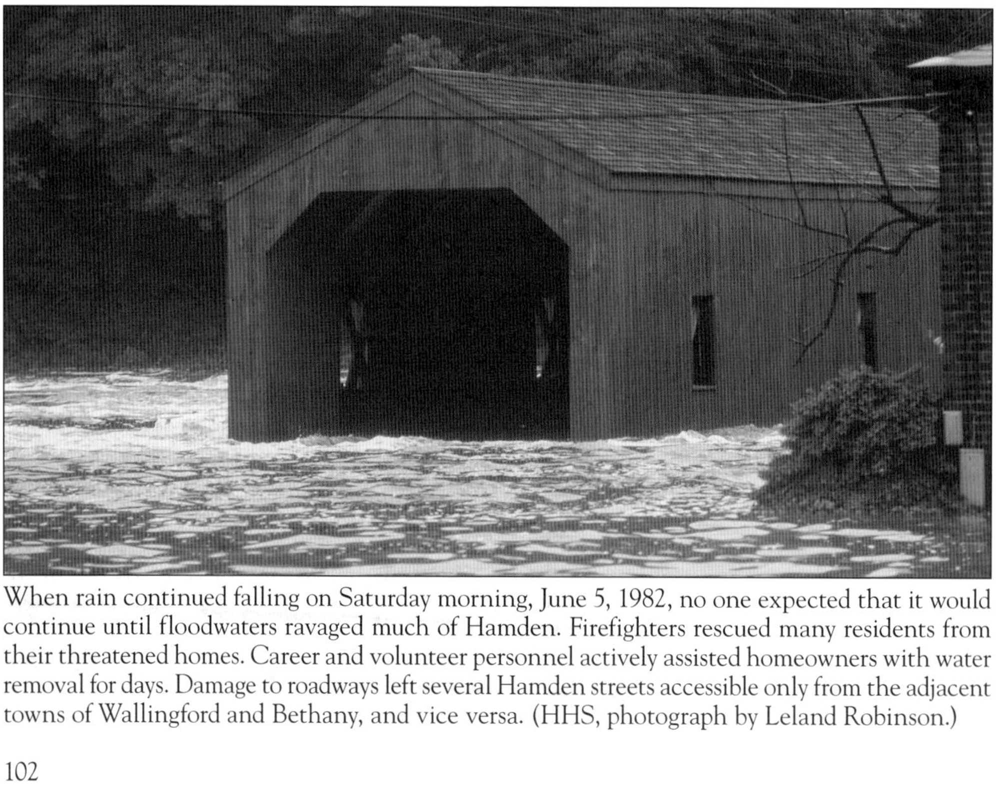

When rain continued falling on Saturday morning, June 5, 1982, no one expected that it would continue until floodwaters ravaged much of Hamden. Firefighters rescued many residents from their threatened homes. Career and volunteer personnel actively assisted homeowners with water removal for days. Damage to roadways left several Hamden streets accessible only from the adjacent towns of Wallingford and Bethany, and vice versa. (HHS, photograph by Leland Robinson.)

Five

Moving Forward

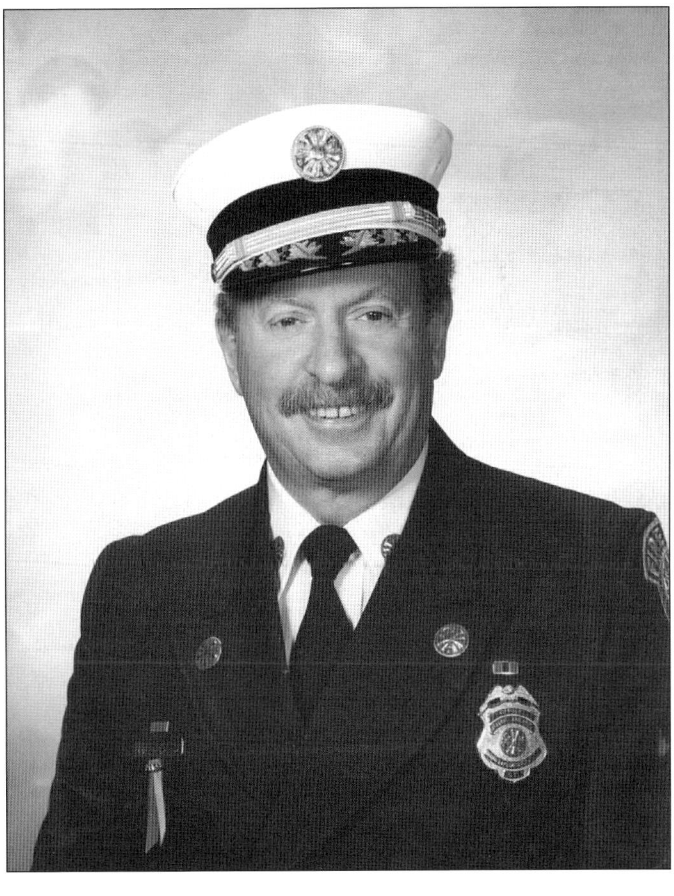

John Tramontano was appointed in January 1959. Promoted to lieutenant in 1973, Tramontano became one of Hamden's first certified paramedics in 1976, the same year he organized the department's Mountain Rescue Team. Appointed deputy chief training officer in 1981, Tramontano began a vigorous training program for all career and volunteer personnel that, when he became chief in 1984, included mandatory firefighter certification. (Fire Chief's Office.)

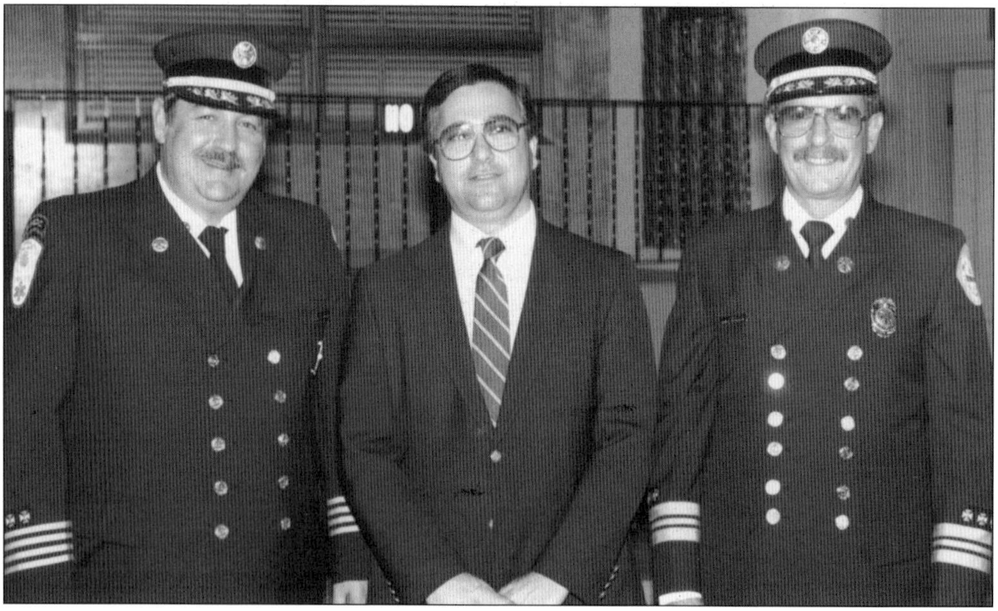

Assistant Chief Walter Macdowall (left), Commissioner Robert LaTorracca, and Chief Tramontano are pictured at the department's 1985 awards ceremony. Like Tramontano, Macdowall was among the department's first certified paramedics. In 1983, he was appointed as the department's first Emergency Medical Services (EMS) Officer, coordinating all EMS matters and related training. Macdowall retired in 1991 to become chief of the LaGrange (Illinois) Fire Department. He passed away in 1995. (Lawrence Gershman.)

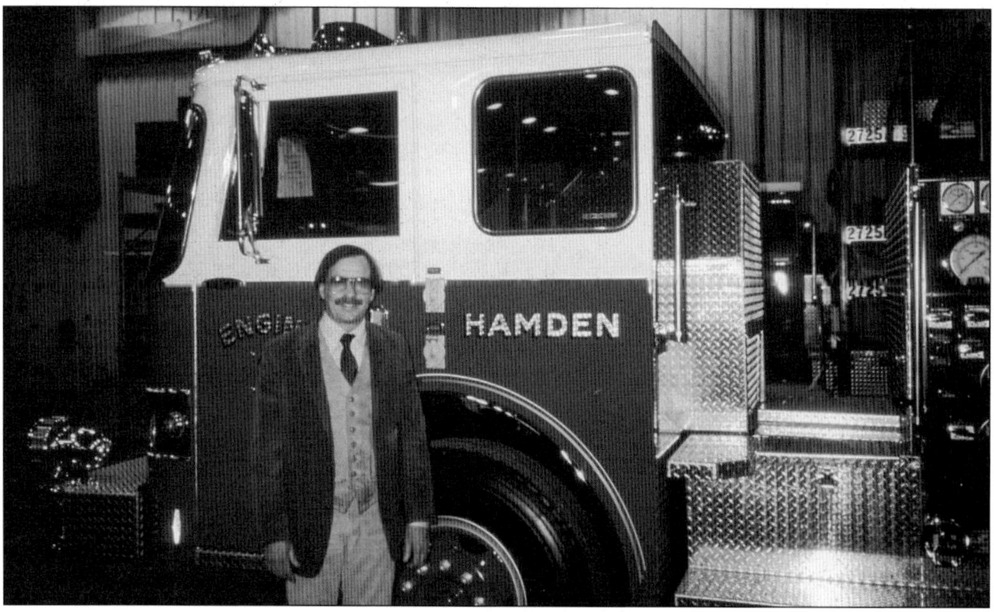

At Pierce's Appleton, Wisconsin, factory in November 1984, Chief Tramontano stands beside one of two Hamden Pierce Dash 1,000-gpm pumpers ordered during Chief Leddy's last year on the job. Tramontano and Asst. Supt. Paul Wetmore Sr. drove the pumpers back to their new home in Hamden. They were Hamden's first new full-size pumpers since 1974. (John Tramontano.)

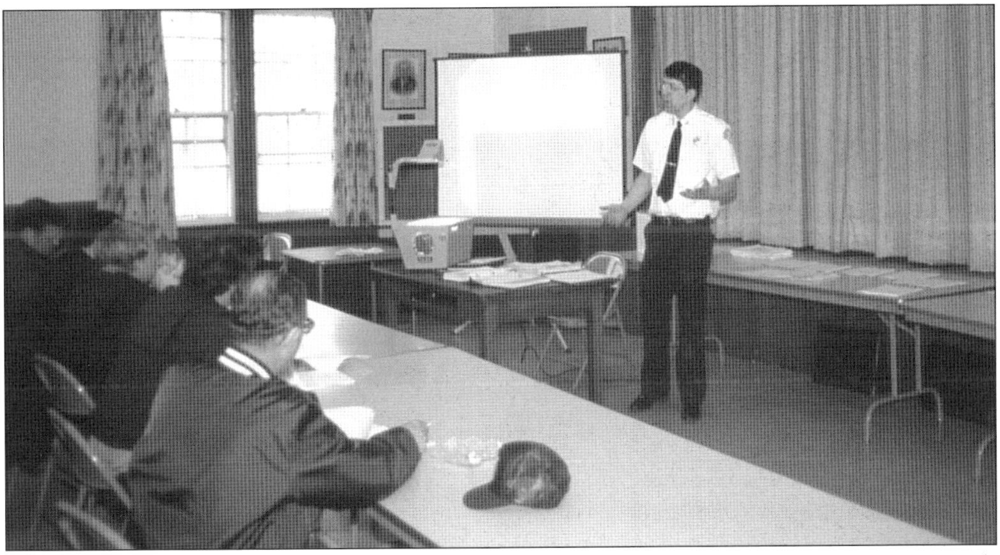

As part of a department reorganization that included adding two more lieutenants to each platoon, Chief Tramontano required state Fire Fighter I certification for all career and volunteer personnel. This class for career firefighters in 1984 was conducted by the author, who served as the department's training officer from 1984 to 1990. (John Tramontano.)

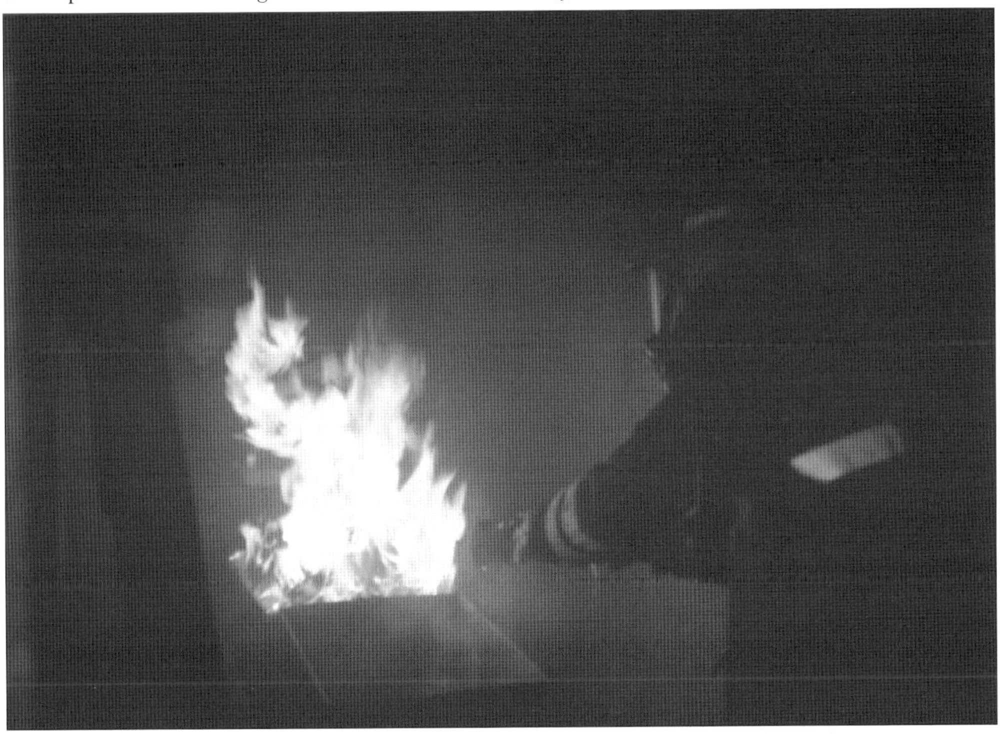

Periodic live firefighting training, even for the most seasoned veterans, reinforces the same training they received as recruits. In the 1980s, the Wallingford Fire Department generously offered the use of its training facilities on Hartford Turnpike to Hamden's training division, which otherwise had to rely on donated derelict Hamden buildings. This December 1984 photograph was taken in the basement of Wallingford training tower. (John Tramontano.)

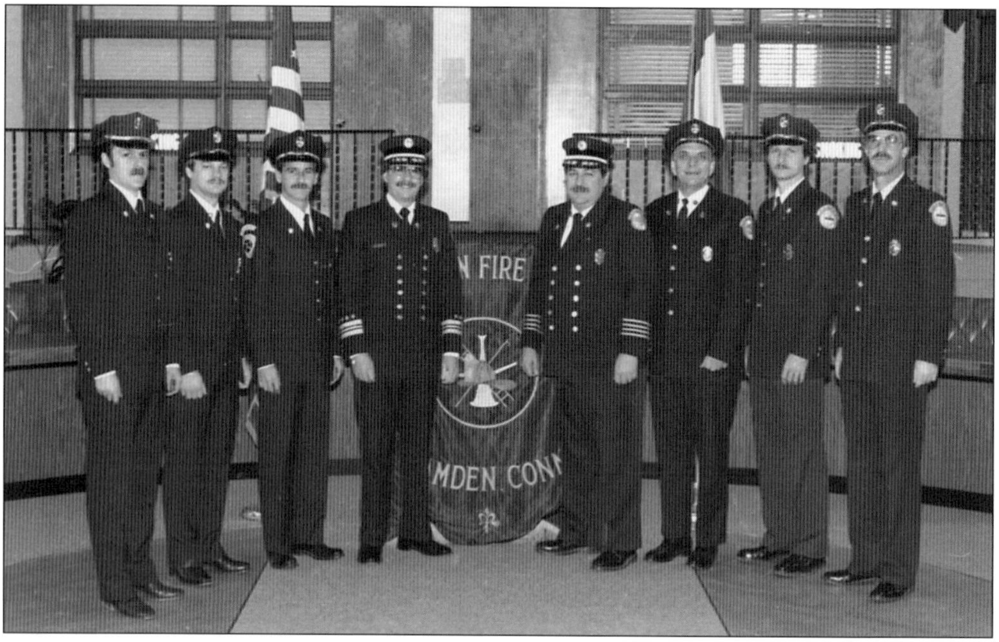

Seven veteran firefighters were promoted to lieutenant in November 1984. Pictured are, from left to right, Lt. Howard Hurlburt Jr., Lt. Dennis Baker, Lt. Robert Viglione, Chief Tramontano, Asst. Chief Macdowall, Lt. Robert Kelo, Lt. David Strawhince, and Lt. Robert Mordecai. Lt. John Calamo was on a call. The chiefs are wearing the more traditional round firefighter's dress cap, which the department was transitioning to at the time. (HFRA.)

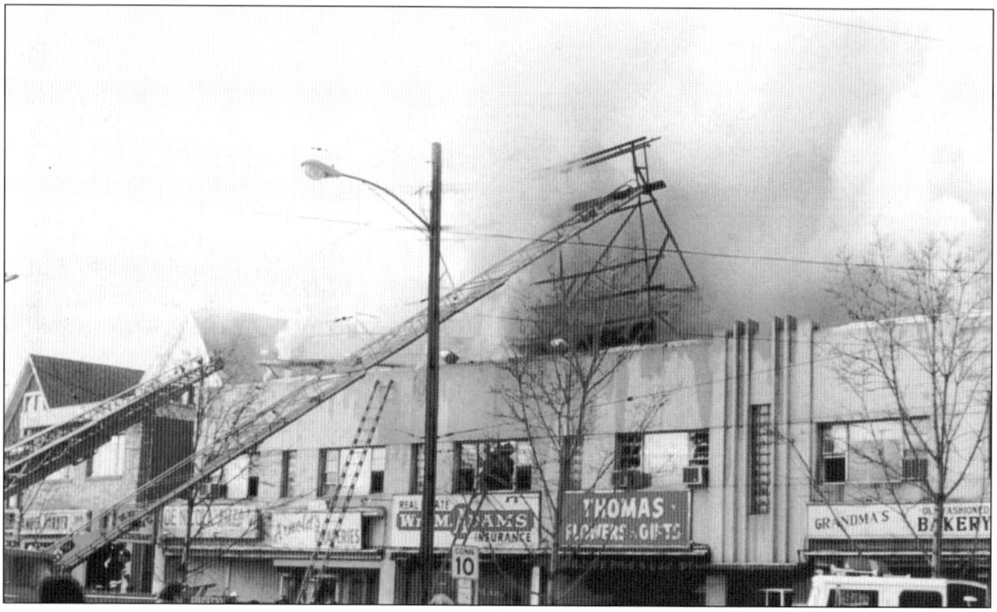

On Friday, January 25, 1985, a fire in an electrical panel extended up through several false ceilings and spread throughout this two-story commercial occupancy at 1206 Dixwell Avenue, popularly known as the "French-Italian Building," Although firefighters' initial efforts to locate and extinguish the fire were hampered by the false ceilings and collapsing ductwork, the 38-year-old structure in Hamden's "Bank Section" was saved and reopened by year's end. (Joseph Ciscone.)

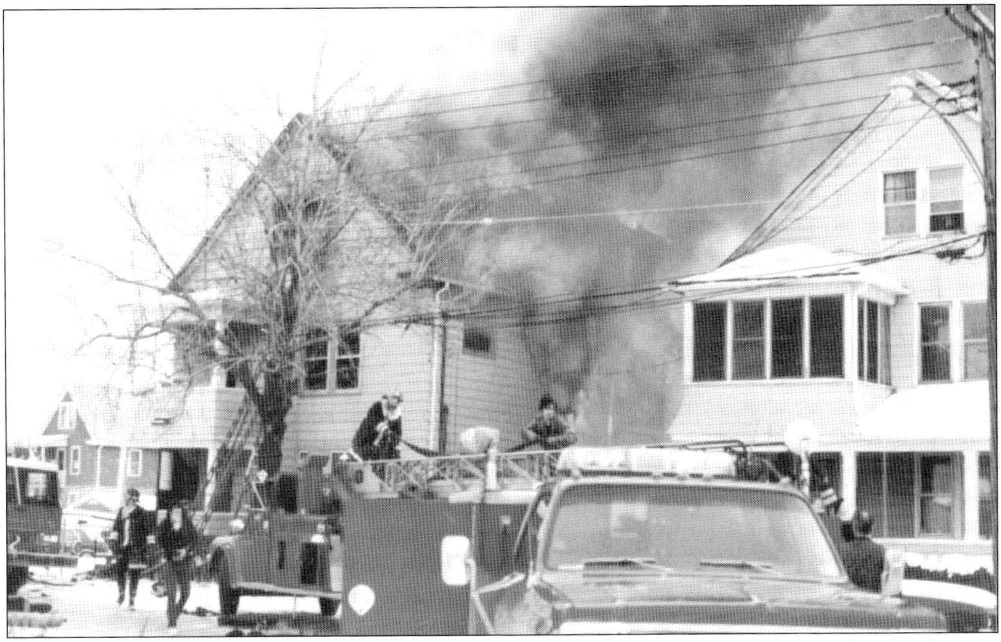

On February 9, 1985, a bitter cold Saturday just two weeks after the French-Italian fire, firefighters were dispatched to 27 Edwards Street for an odor of smoke. On arrival, the lone engine company reported fire on the top floor, whereupon the dispatcher filled out the box assignment. A 23-year old pregnant woman was trapped and unable to escape. Firefighters forced entry to her apartment and brought her safely outside. (Joseph Ciscone.)

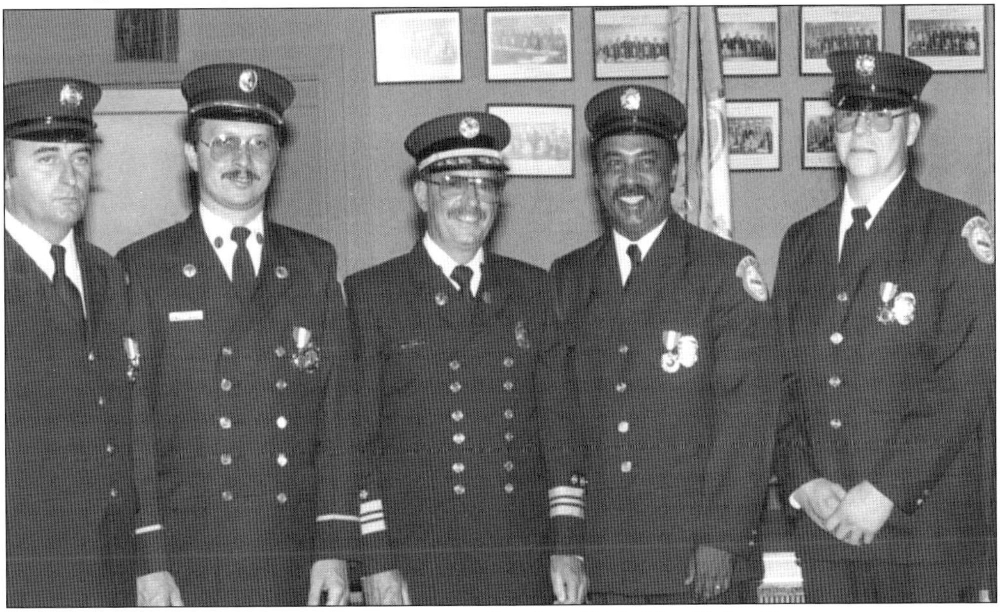

At the department's first awards ceremony in October 1985, Chief John Tramontano presented the department's Medal of Valor to the Engine 2 personnel who rescued the 23-year-old pregnant woman from her burning Edwards Street apartment the previous February. From left to right are Firefighter James Moore, Lt. Donald LaBanca, Chief Tramontano, and Firefighters Wayne Lowry and Carl Backus. (John Mongillo Jr.)

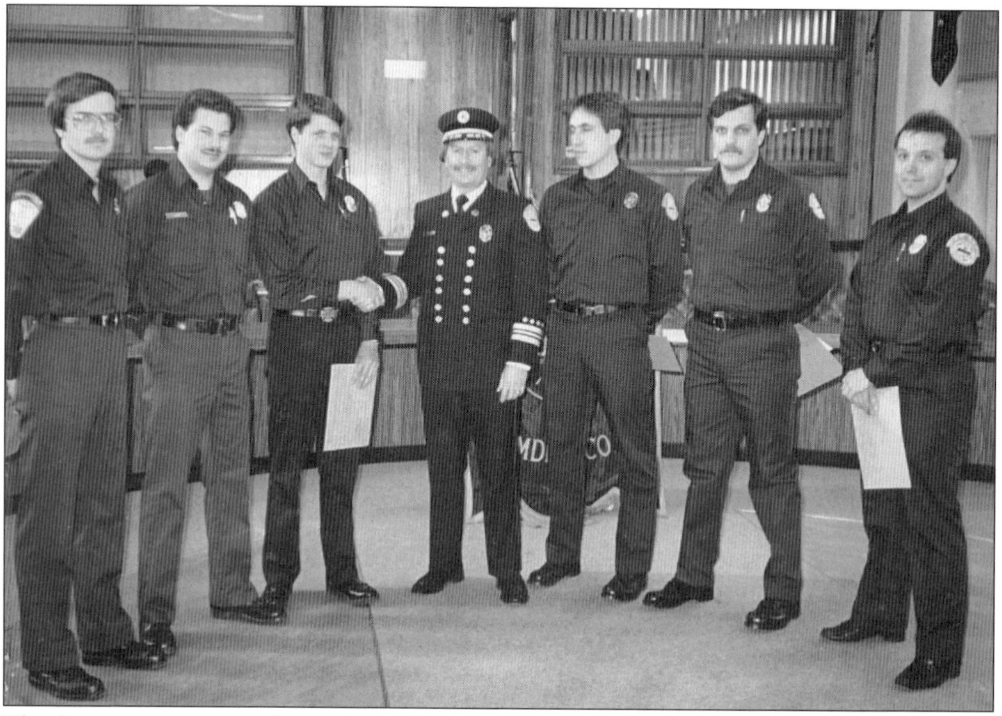

The first new recruits in five years were appointed in February 1987. Following completion of department training, which included Fire Fighter I certification, the new firefighters received their badges at a ceremony in the town hall auditorium. From left to right are Firefighters Dennis Harrison, William Fitzmaurice, John Spencer, Chief Tramontano, and Firefighters Ronald Desroches, Timothy Sullivan, and Kevin O'Hare. (HFRA.)

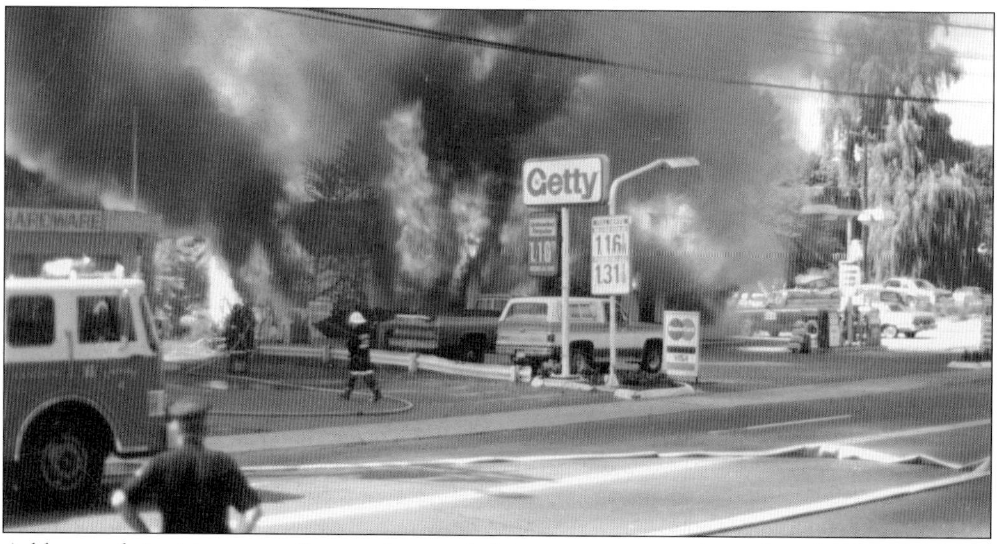

A blowtorch igniting a gasoline tank caused this service station fire only a few hundred feet from Mt. Carmel Station 5. Arriving on scene in less than two minutes, Engine 5's two-man crew determined there was no life hazard and began protecting the endangered building next door. Once additional companies arrived, crews concentrated on the main body of fire, which was extinguished in short order. (Clark Hurlburt.)

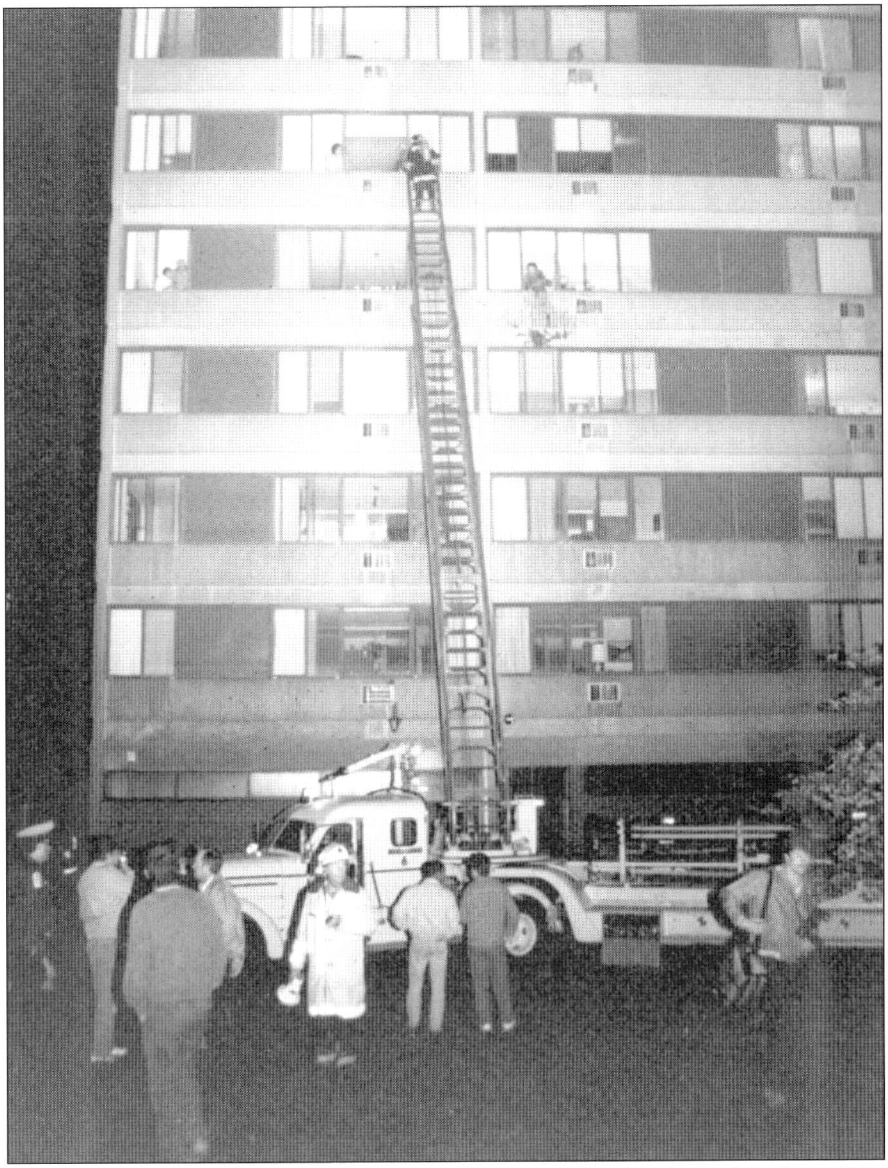

Department personnel and resources were pushed to the limit on the night of May 25, 1988. Davenport Residence at 125 Putnam Avenue, an apartment building for elderly residents built in 1970, had no sprinkler system. When fire broke out in the first-floor library, heavy smoke quickly traveled throughout the seven-story structure, requiring a rapid and systematic evacuation of every resident from the building's 214 apartments. All on-duty and numerous off-duty career and volunteer personnel responded. Mutual aid was requested from New Haven, Cheshire, and Bethany Fire Departments. Chief John Tramontano (in white) is pictured with New Haven's Truck 6, which assisted in evacuating residents from the Clifford Street side of the building. Cheshire responded with its bucket-equipped Snorkel, an engine, rescue, and a unit to fill tanks. C-MED's "disaster plan" was activated, bringing emergency medical personnel from Cheshire, North Haven, Bethany, New Haven, and all available ambulances. A quick and aggressive interior fire attack and prompt requests for mutual aid resulted in limited damage, a successful evacuation of all residents, and effective triaging and transport of the injured. (Tara Tramontano.)

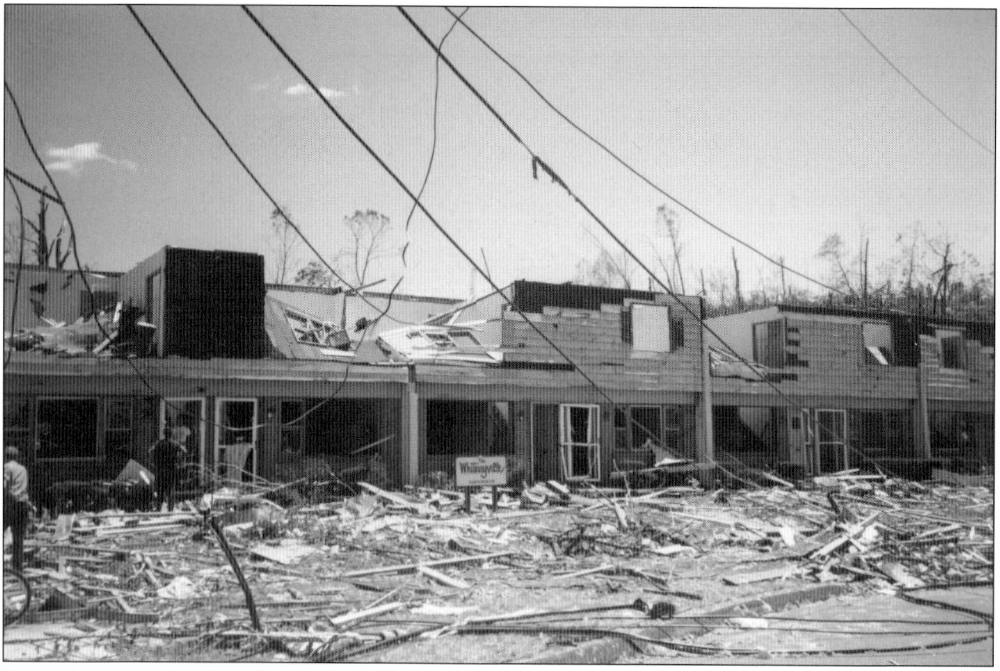

Tornadoes do occur in Connecticut. This was all that was left of a row of condos on Newhall Street when an F4 tornado struck the Whitneyville-Newhall section of Hamden about 5:00 p.m. on July 10, 1989. Winds reaching up to 200 miles per hour left a swath of destruction resulting in millions of dollars in damages. (Edward Doiron Jr.)

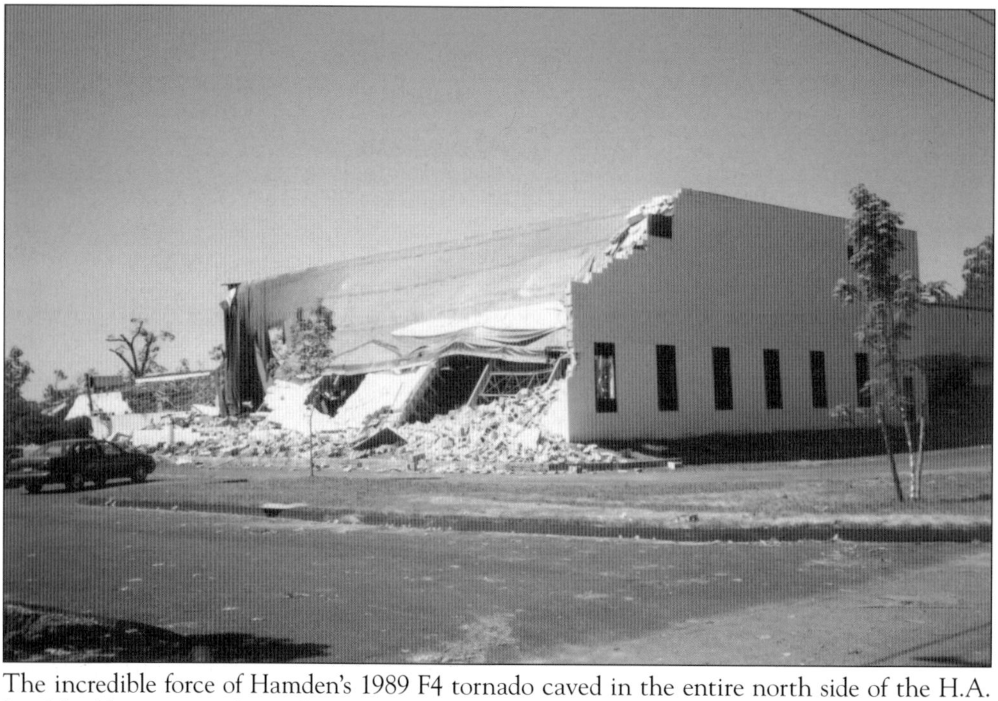

The incredible force of Hamden's 1989 F4 tornado caved in the entire north side of the H.A. Leed building on Leeder Hill Drive. Scores of other structures in the area suffered similar or worse damage. (John Corbett.)

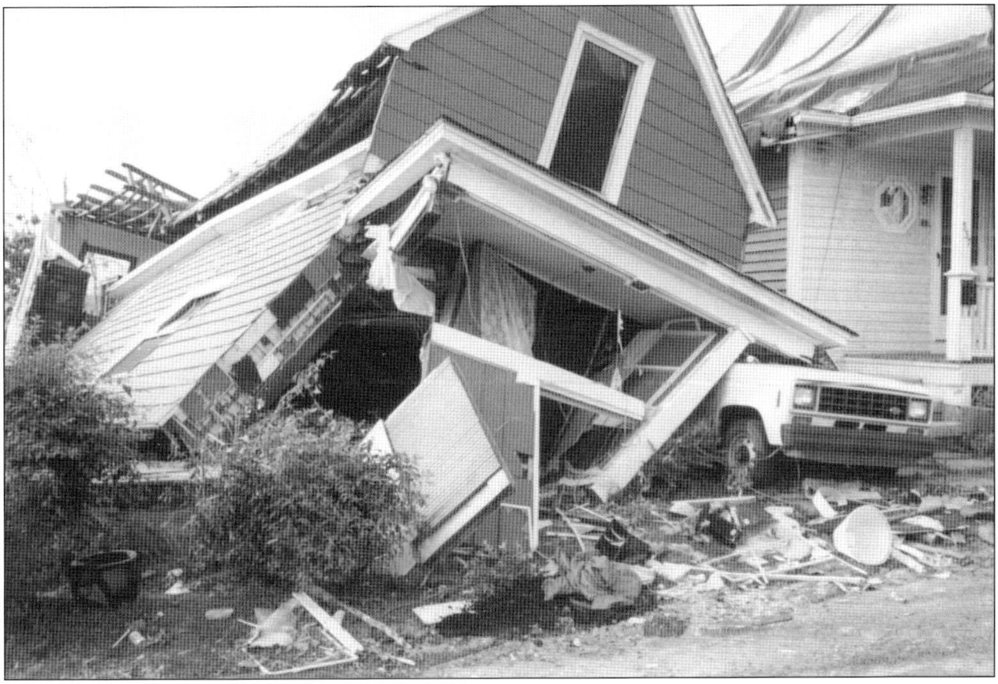

Auger Street resident Harold Wheatley shot this photograph of a devastated Newhall Street house, which was typical of the extreme damage inflicted by the tornado within the Dixwell Avenue, Putnam Avenue, Whitney Avenue, and Goodrich Street area. (Harold Wheatley.)

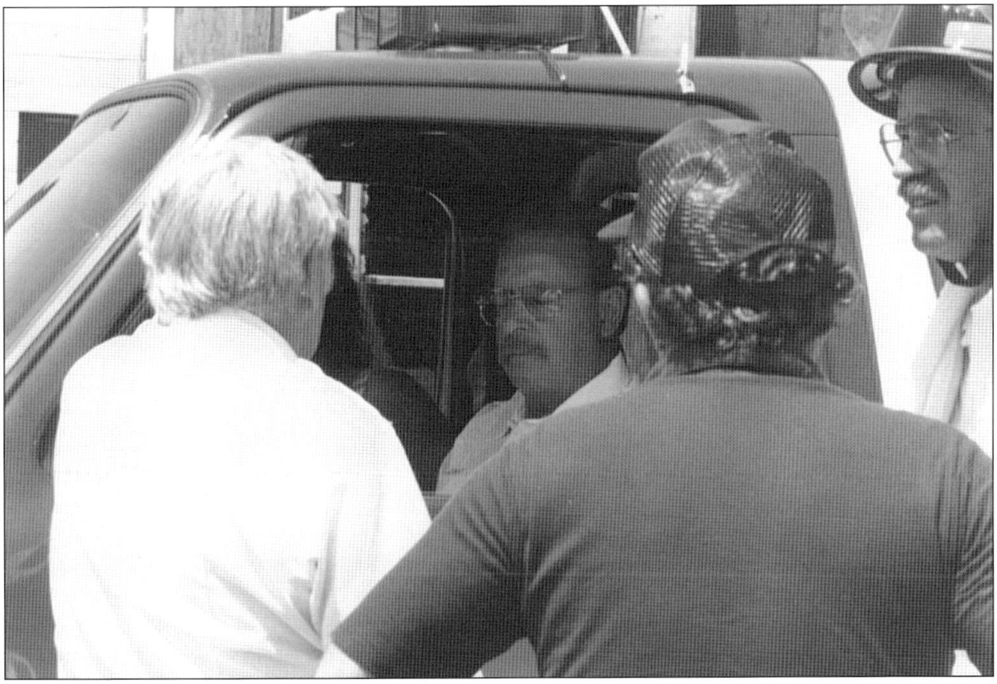

Standing beside Car 1 are, from left to right, Commander John Laffin, Firefighter John Corbett, and Lt. Robert Mordecai, receiving instructions from Chief John Tramontano, in Car 1, who coordinated department operations during the days immediately after the 1989 tornado. (Raymond Dobbs.)

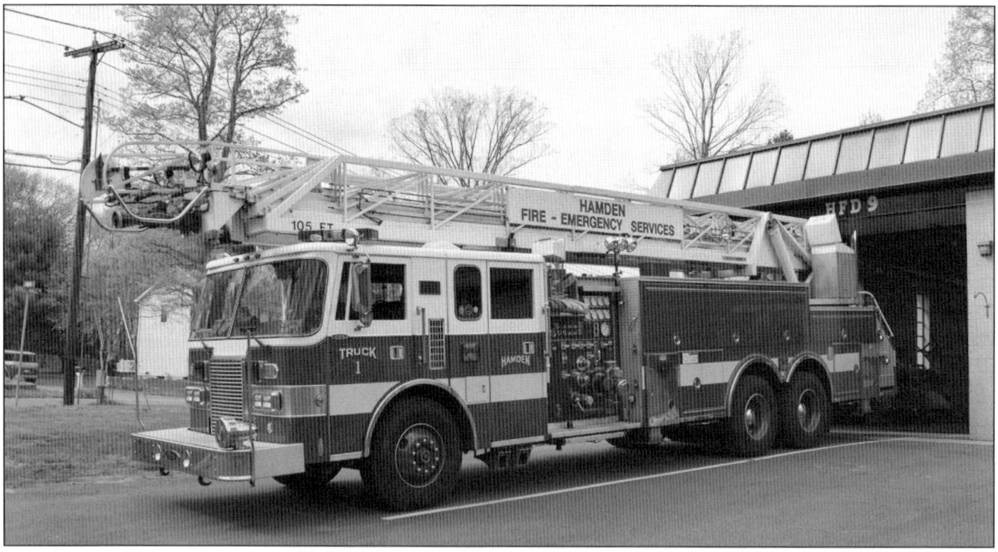

In 1985, both two-man ladder companies were consolidated into a single four-man company, assigned to the aging 1970 Maxim 100-foot aerial truck, which was redesignated "Truck 1." The 1958 Maxim aerial truck, now "Truck 2," became a spare. In 1989, the 1970 Maxim's aerial was inspected and deemed unsafe. East Haven lent Hamden a spare aerial truck while a new Hamden aerial ladder truck was on order. In fact, the town had two on order. The first to arrive was a 1990 Pierce 105-foot rear-mount aerial with a pump (above)–a "quint." It went in service at Station 3, replacing the loaner. A Pierce 100-foot rear-mount tower ladder (below) followed in 1991, replacing the quint, which went to Station 5 as Truck 5. Truck 5 could run as either an engine or truck company. (Both, HFRA.)

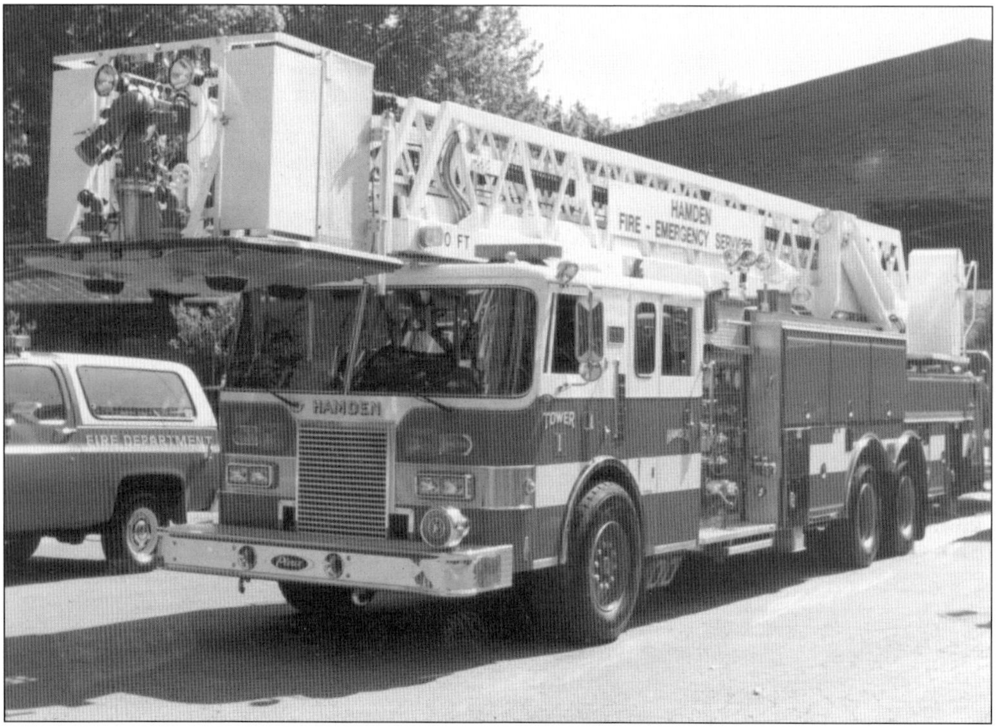

Firefighter Ronald "Ronnie" Altieri joined the department in 1967. A definite firehouse standout, Altieri was a dedicated firefighter who was blessed with a wonderfully sharp, irreverent, and often self-deprecating sense of humor. In early 1991, Ronnie was stricken at the fire station and hospitalized. While recovering, Firefighter Altieri died suddenly on May 2 of that year, leaving behind his wife, Diane, and two children. (John Tramontano.)

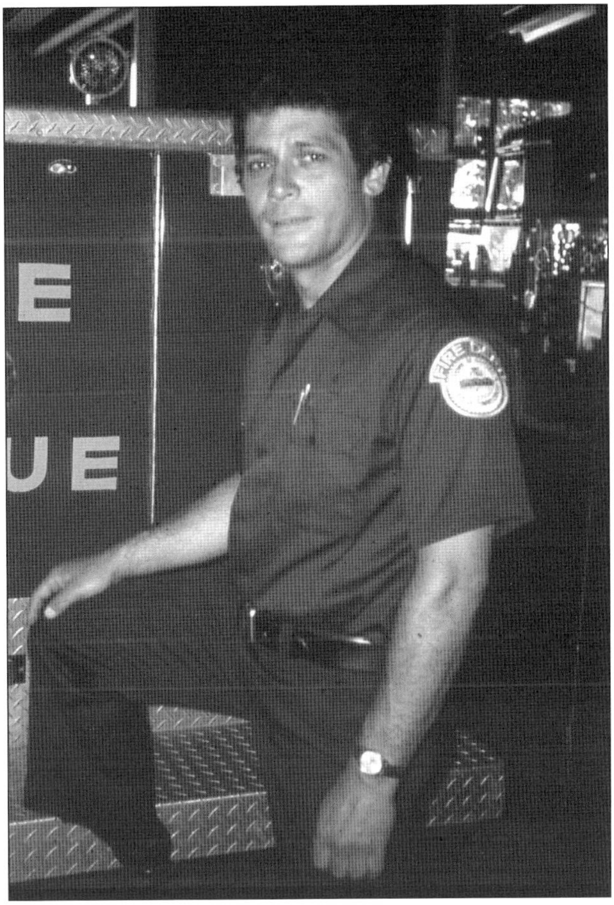

A Hamden tradition began in 1993, when members of the Mt. Carmel, Mix District, and Dunbar Hill companies sponsored Hamden's first Volunteer Firefighters Fireworks at Town Center Park. For over two decades, Hamden businesses, service clubs, and residents have helped finance this Independence Day event that is among the best in southern Connecticut. Pictured here in 1999 is Mix District's Duane Wetmore, an early committee chairman. (HFRA.)

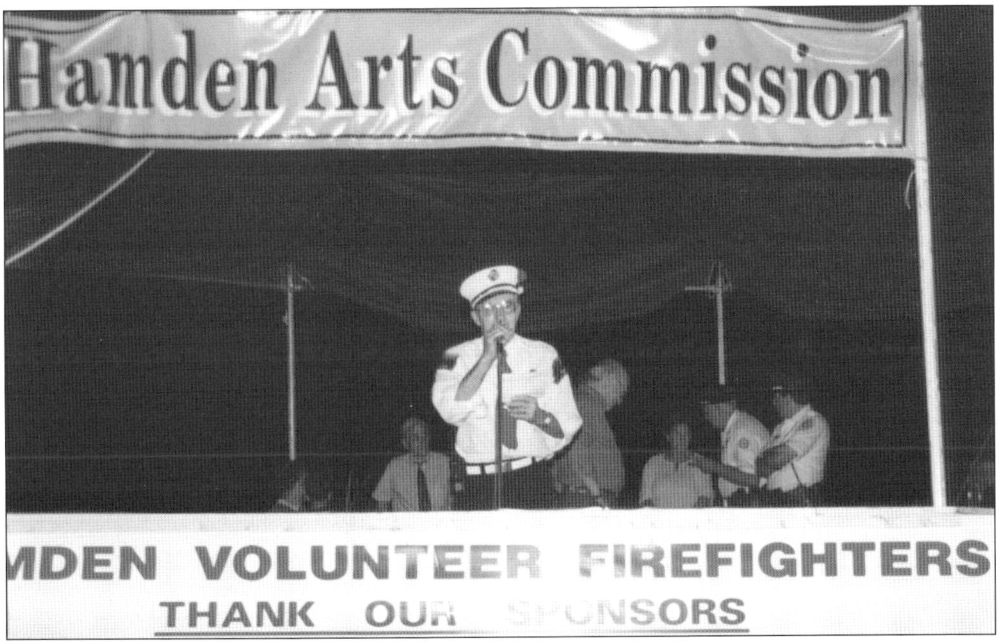

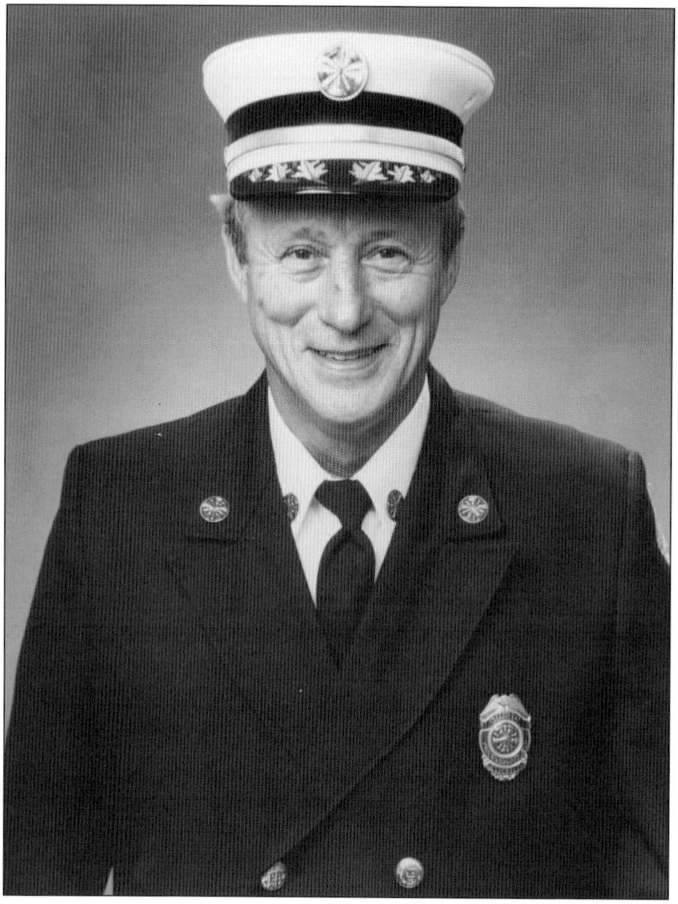

Paul Wetmore Sr., a member of a large and active Hamden firefighting family, began his fire service career as a youth at Mix District Volunteer Fire Co. 7. Joining the career department in 1962, he became assistant superintendent of alarms and apparatus in 1969 and superintendent in 1986. Wetmore was appointed department's second-in-command in 1991, and fire chief in 1993. (Fire Chief's Office.)

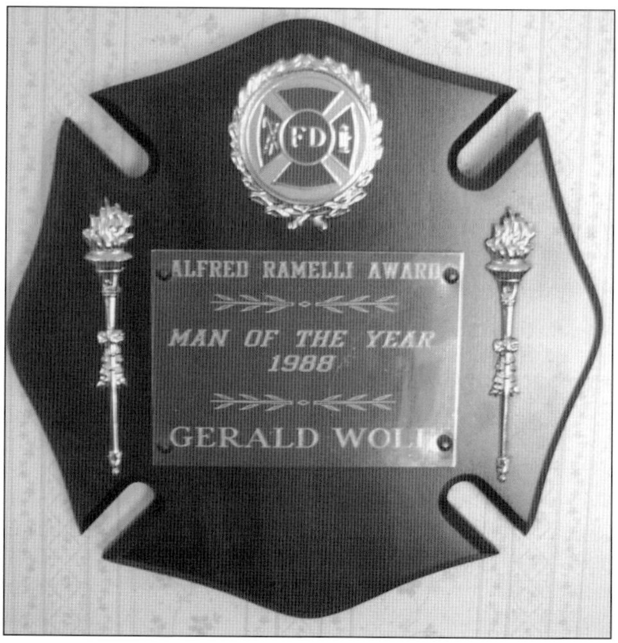

For 25 years following Firefighter Alfred Ramelli's untimely passing, the Alfred Ramelli Award for Man of the Year was presented each year to a department member who best represented the friendly "everyman" spirit of Firefighter Ramelli, a man respected and appreciated by his fellow firefighters. In 1988, Firefighter Gerald Wolf was awarded the honor by his peers. (David Johnson.)

Hamden hosted Team Australia during the 1995 Special Olympics World Games, the year's largest organized athletic event. Pictured in the first row in front of Tower 1, along with other members of Team Australia, are, from left to right, Battalion Chief Michael Ambriscoe, Chief of Department Paul Wetmore Sr., an unidentified Australian athlete, and Battalion Chief Thomas Doherty. Doherty and Ambriscoe coordinated the department's fire, emergency medical, and logistical preparations for the 10 days of the games. (HFRA.)

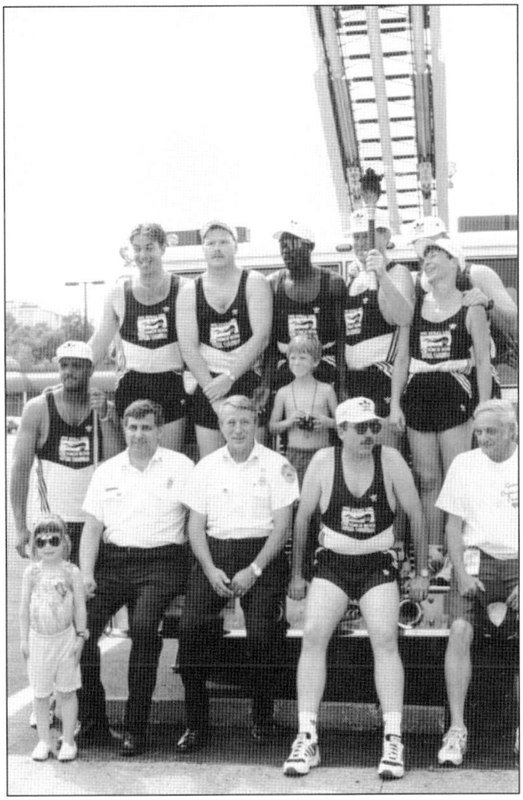

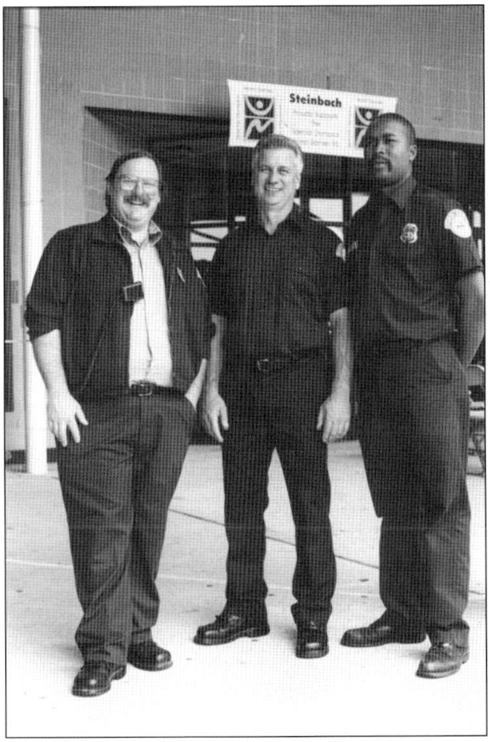

Pictured from left to right, Lt. Clark Hurlburt and Firefighters Robert Anthony and Joseph Anderson were assigned to "temporary" Engine 12 at the Hamden Mart during the World Games bowling competition. With three Hamden venues, department members spent over a year preparing for additional emergency incidents due to the anticipated large influx of visitors to Hamden. Local 2687 members and many out of town firefighters donated thousands of hours to the effort. (HFRA.)

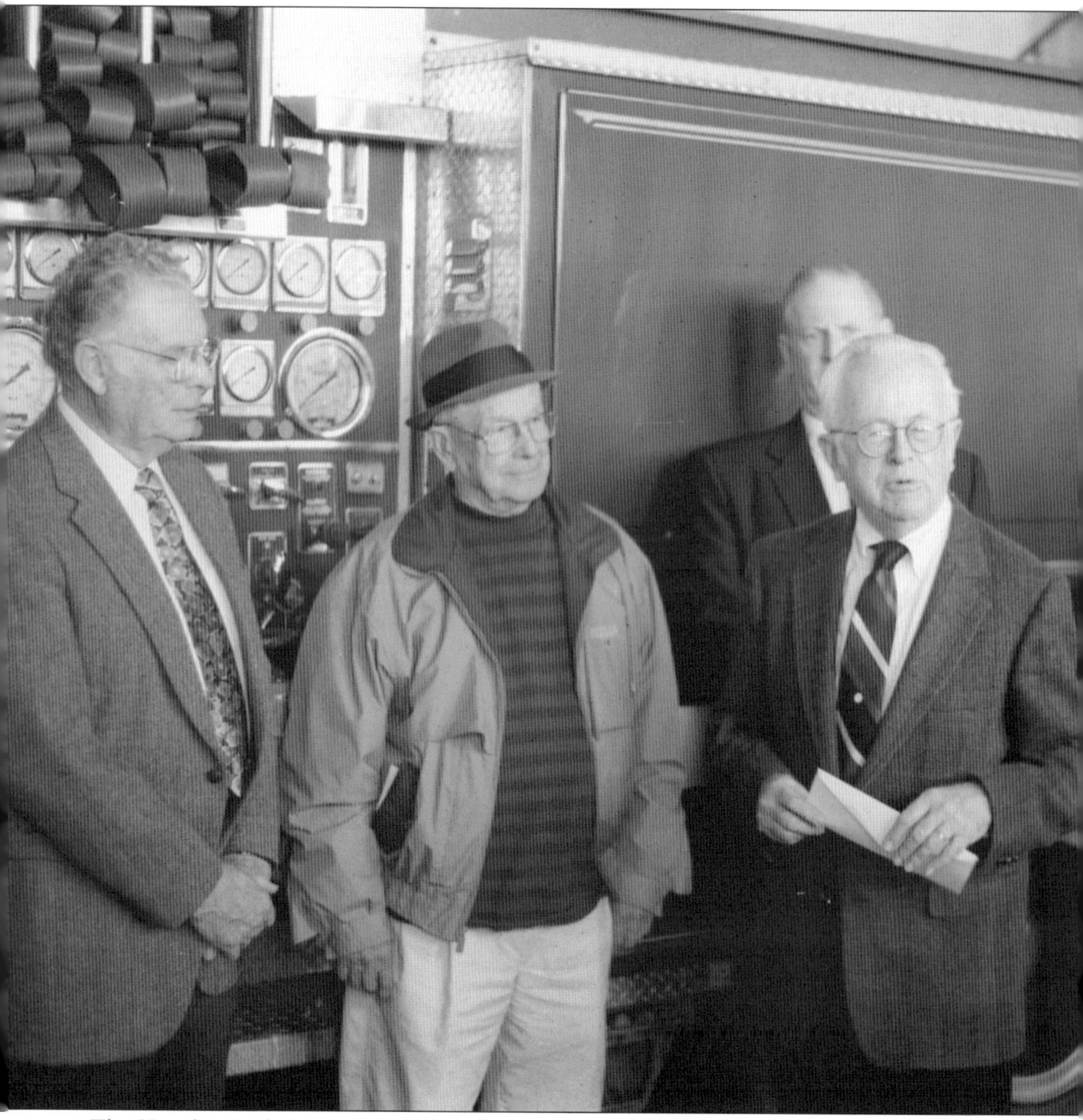

The Hamden Paid Firemen's Sick Benefit Association (HPFSBA) was the bargaining unit for Hamden's firefighters until it was replaced in 1978 by an interim bargaining unit, the Hamden Firefighters Association. After the Hamden Professional Firefighters Local 2687 IAFF was organized the following year, the HPFSBA remained in existence to assist its remaining members with minor medical expenses. By the early 1990s, membership in the HPFSBA had dropped to a handful of members. In November 1995, the remaining members voted to dissolve the association and donate its substantial treasury to Connecticut Hospice. On December 22, 1995, in a brief ceremony at Centerville Station 4, two of the association's founding members, retired Deputy Chiefs George Reutenauer (left) and Francis Leddy, presented a check for $15,484.50 to Rev. John Abbott of Connecticut Hospice. Leddy served as the association's first secretary when it was organized in 1948. Standing behind Reverend Abbott is Battalion Chief Gilbert Spencer (retired), the association's last president. (David Johnson.)

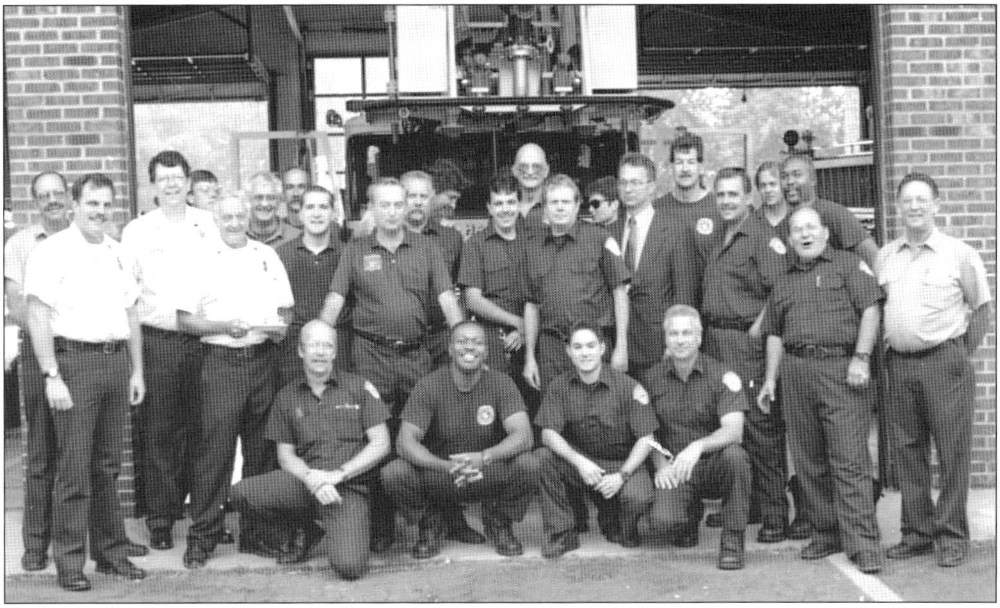

On August 29, 1996, Platoon 3 personnel said farewell to their shift commander, Battalion Chief Thomas Doherty (third white shirt from the left), who was retiring after 35 years on the job. Starting as a volunteer with Mt. Carmel, Doherty was appointed a career firefighter in 1961 and rose to battalion chief by 1984. His late father, Deputy Chief Everett Doherty, retired in 1966 after 39 years. (HFRA.)

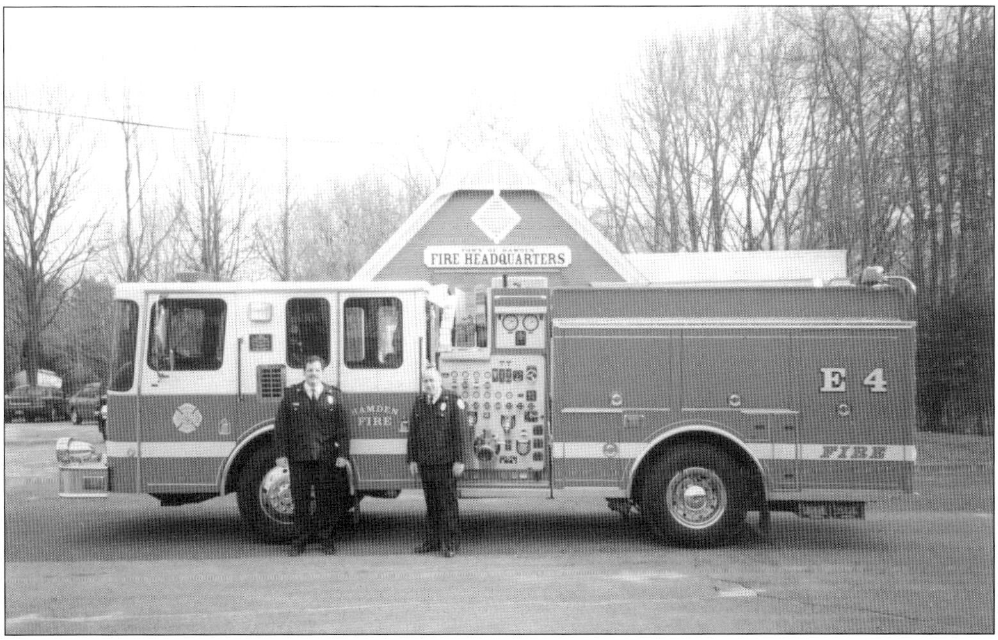

A new piece of fire apparatus is an exciting event for every fire department. In 1999, Chief Timothy Sullivan (left) and Deputy Chief Edward Badamo received delivery of this new 3-D 1,250-gpm pumper, equipped with a 750-gallon tank and an internal Class-A foam system. The building in the background, once Hamden's last one-room schoolhouse, served as fire headquarters for about a decade. (Timothy Sullivan.)

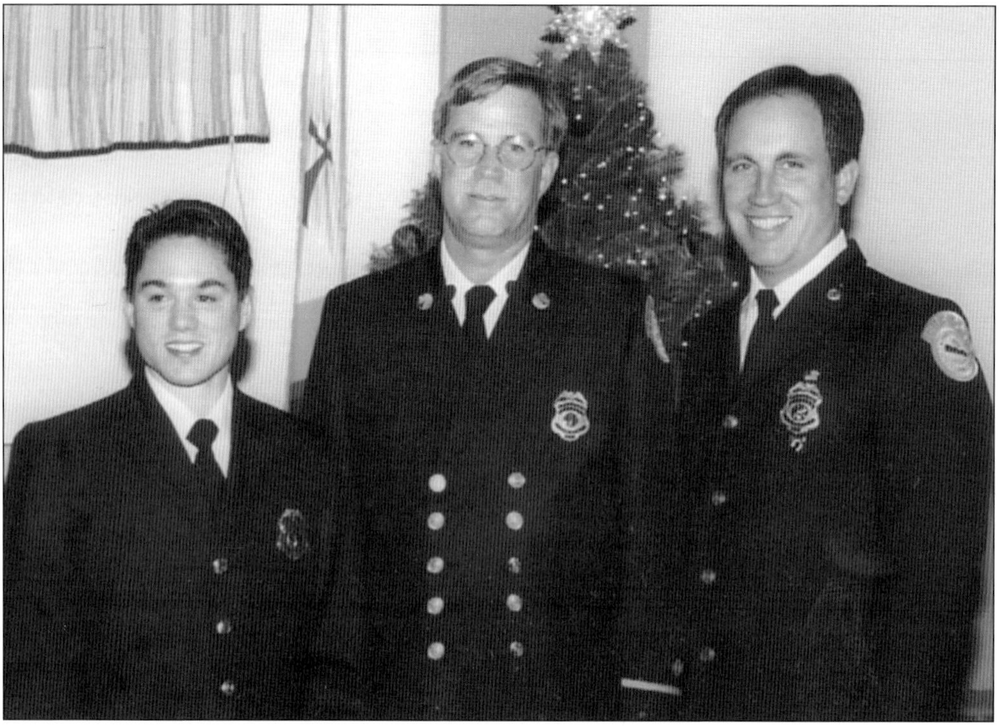

From left to right, Firefighter Henry Puciato, Lt. Robert Mordecai, and Firefighter Ronald Desroches were recognized by the Knights of Columbus in 1998 for their remarkable rescue of an unconscious woman from inside a fully involved apartment on Mix Avenue. Commendations to the three were presented by Hamden mayor Barbara DeNicola. (HFRA.)

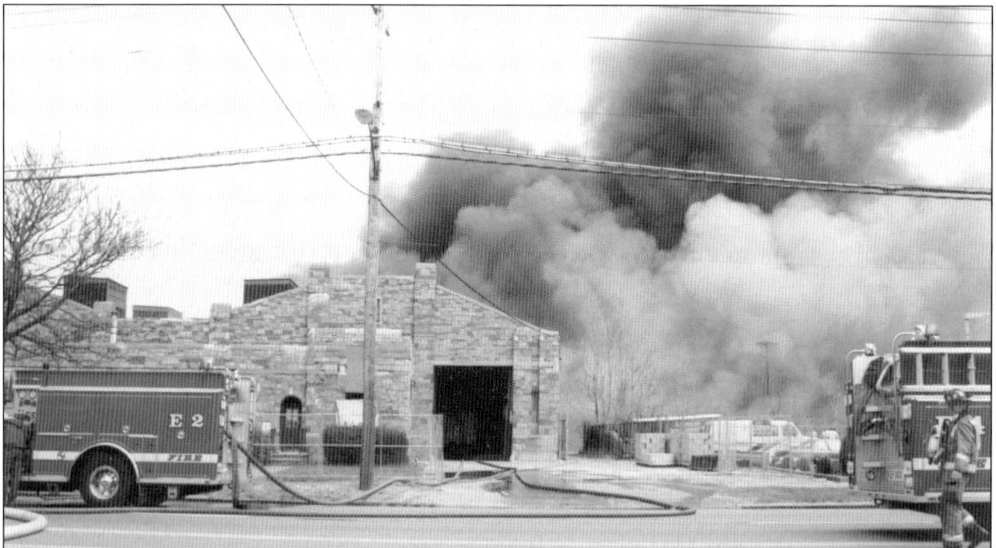

A major Hamden industry was the High Standard Manufacturing Corporation at 1817 Dixwell Avenue, producer of .50-caliber machine guns for Allied aircraft during World War II. A workman's power saw cutting pipes in the 30,000-square-foot vacant building sparked this three-alarm fire on April 29, 1999. This is presently the site of Hamden's Home Depot. (Jeffrey Pechmann.)

Six

A New Century

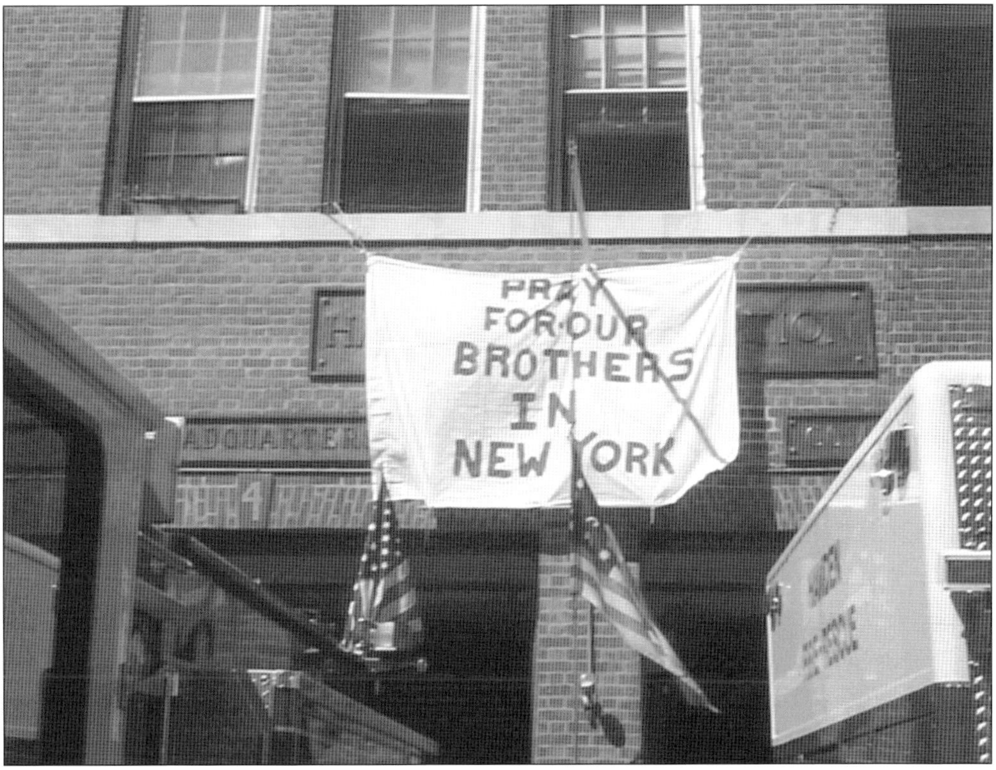

On September 11, 2001, the country experienced a terrible, tragic, and vile attack. With 343 New York City Fire Department (FDNY) firefighters among the thousands lost in the collapse of the twin towers of the World Trade Center, 9/11 has become a defining moment for the fire service and firefighters everywhere and, sadly, an ominous warning in the new century. (HFRA.)

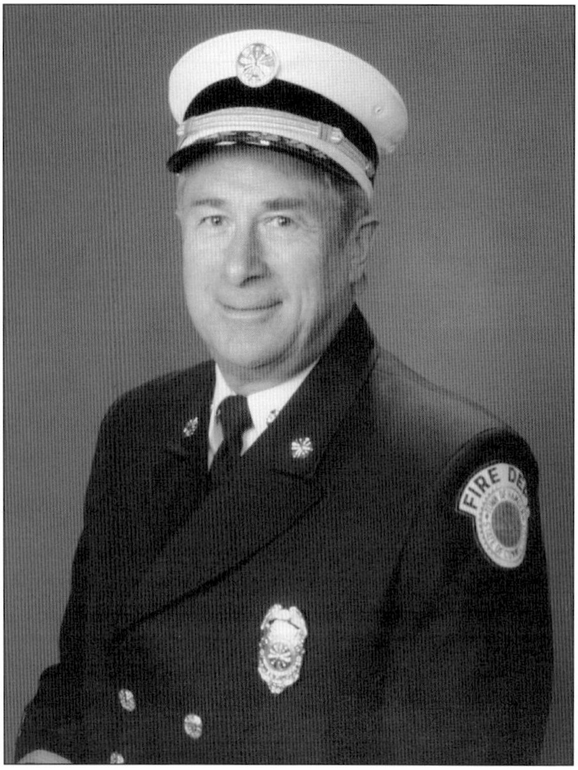

Chief James Leddy joined the department in 1973, advancing through the ranks to lead the department in 2002. In 1982, then lieutenant Leddy and his Platoon 3 crew were commended for successfully rescuing a stranded family during the June flood. Leddy's father was Deputy Chief Francis "Chalky" Leddy. His uncle V. Paul Leddy was the department's longest-serving chief. Leddy's son Brian is a career Hamden firefighter. (Fire Chief's Office.)

The department trains for cold-water and ice rescues. In this February 2002 photograph of an ice rescue training session at Mount Carmel's Clark's Pond, Firefighters John Longo (left) and Seth Patrick are hauled back to land in their wet suits by Firefighter Jason Blythe (left) and Battalion Chief Paul Wetmore Jr. (Robert Mordecai.)

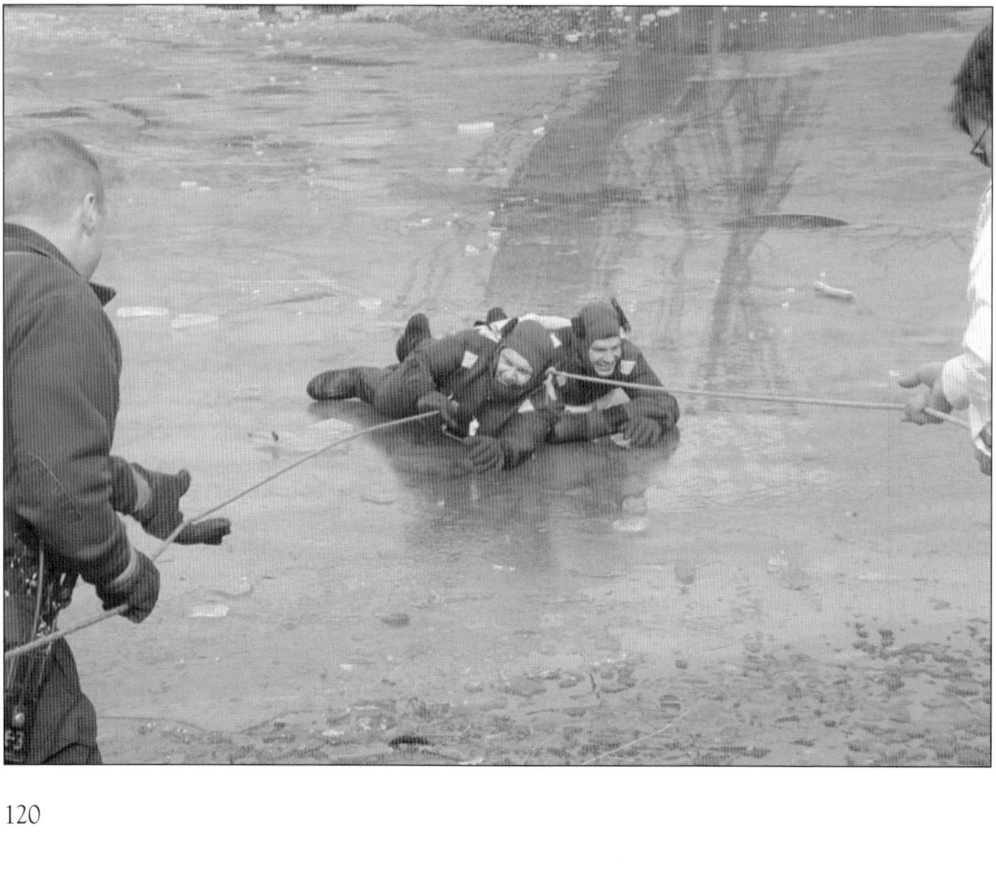

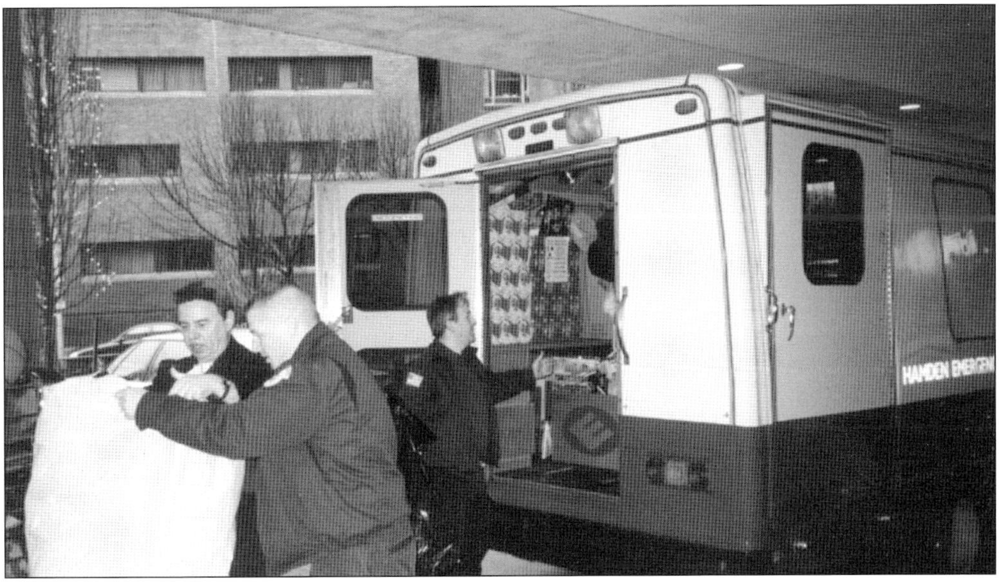

In this 2003 photograph at Yale New Haven Hospital, Firefighters Austin Williams (left), Jason Blythe, and Ralph DiFonzo unload a truck full of new toys collected annually by the Hamden Professional Firefighters Local 2687 IAFF for hospitalized children during the Christmas season. (Robert Mordecai.)

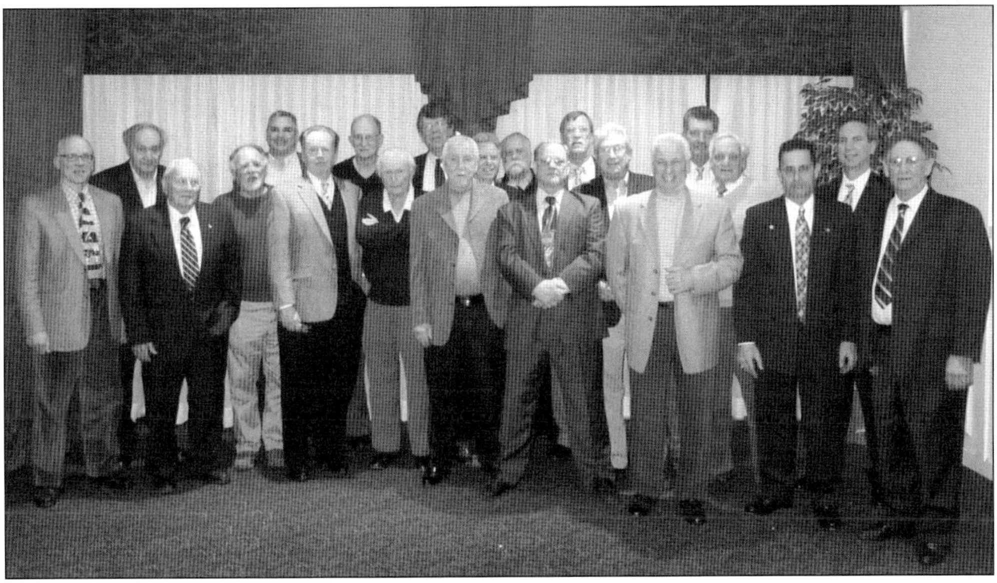

Pictured here at the 2010 Local 2687 Annual Dinner are the some of the nearly 80 members of the Hamden Fire Retirees Association, organized in 2009 to keep Hamden Fire Department retirees connected socially and fraternally, and to preserve the history of the Hamden Fire Department. Members meet quarterly in the Hamden Elks Hall. The association's website is www.hamdenfireretirees.org. (Robert Mordecai.)

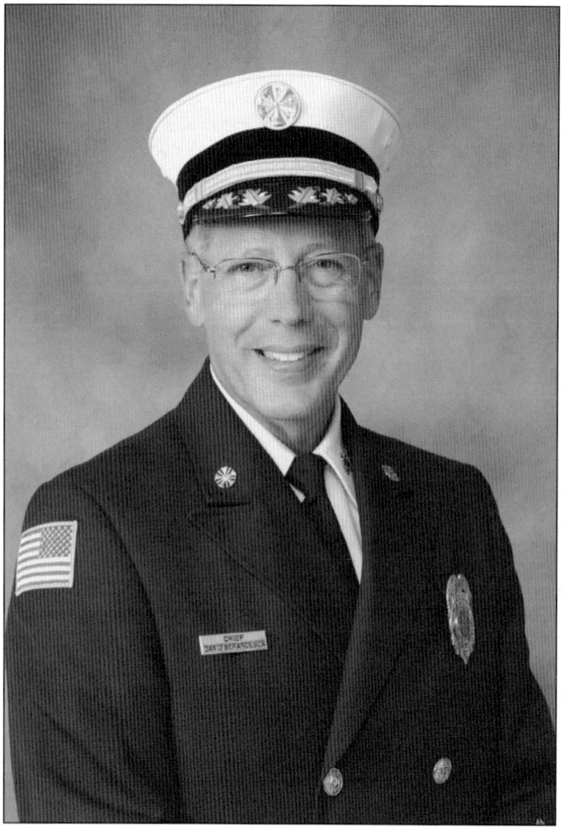

David A. Berardesca began his fire service career in 1977 as a Wallingford volunteer firefighter. Appointed a career firefighter in 1979, he was promoted to lieutenant in 1990 and captain/shift commander in 1995. In 2000, he became Wallingford's assistant chief in charge of fire training and medical services. Following his 29 years with Wallingford, Chief Berardesca was appointed to lead the Hamden Fire Department in 2006. (Fire Chief's Office.)

Deputy Chief Robert Surprise (left) is pictured with Mayor Scott Jackson and Chief Berardesca following his 2010 swearing-in. He succeeded recently retired Deputy Chief Clark Hurlburt. Appointed in 1981 as a firefighter-paramedic, Surprise was promoted to lieutenant in 1988, captain in 1999, and battalion chief in 2006. Deputy Chief Surprise retired in 2013 after serving 32 years with the department. (Robert Mordecai.)

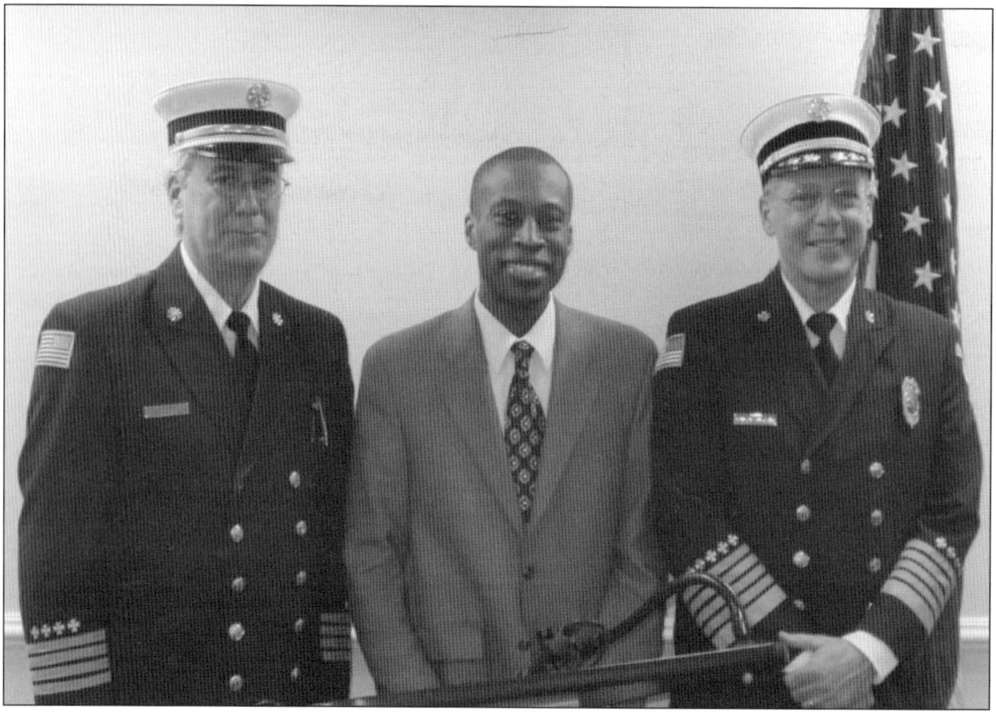

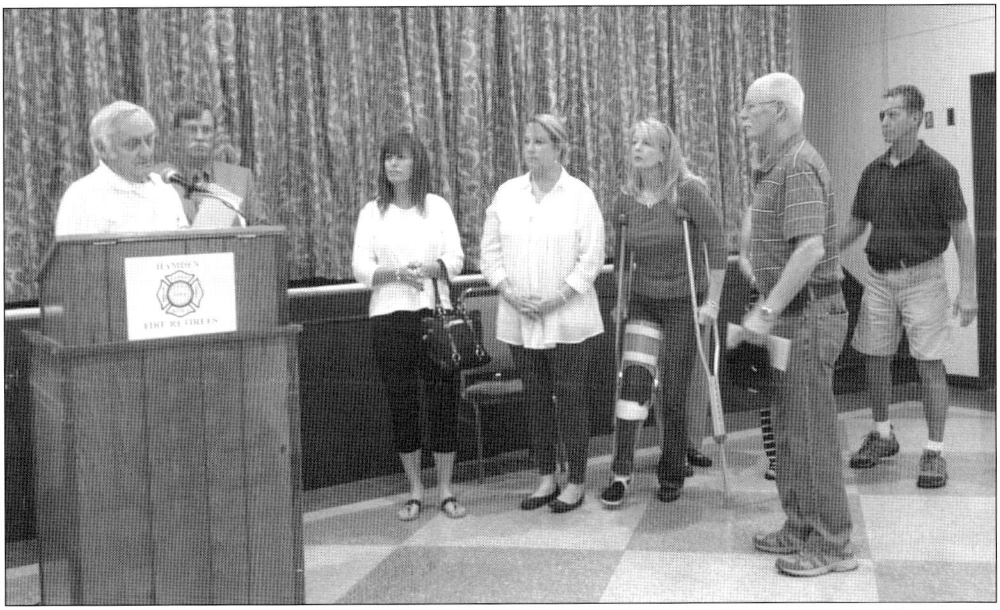

Every October, the Hamden Fire Retirees Association honors members who have passed during the previous year. In 2014, Battalion Chief Thomas Doherty (retired), far left, presented a memorial plaque to the family of Firefighter Paul Reutenauer, who worked on Doherty's platoon. To the right of Doherty are, from left to right, HFRA president Robert Mordecai, Molly McKenna, Sheila DeChello, Kate Jensen, retired Hamden police officer Richard Reutenauer, and Daniel Reutenauer. (David Johnson.)

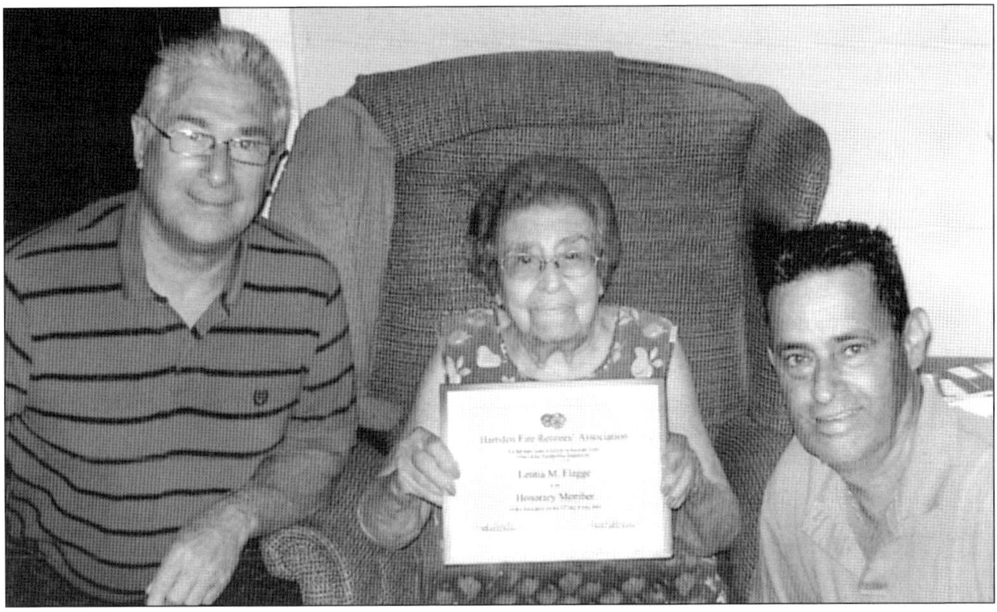

Letitia Flagge, the department's first civilian employee, served as department secretary from 1957 until her retirement in 1978. In 2011, the members of the Hamden Fire Retirees Association voted Flagge to honorary membership in the association. Both retired at the time, Chief James Leddy and Fire Marshal Mark Barletta, HFRA president, presented Flagge with her certificate of honorary membership. Flagge passed away in 2014 at age 100. (David Johnson.)

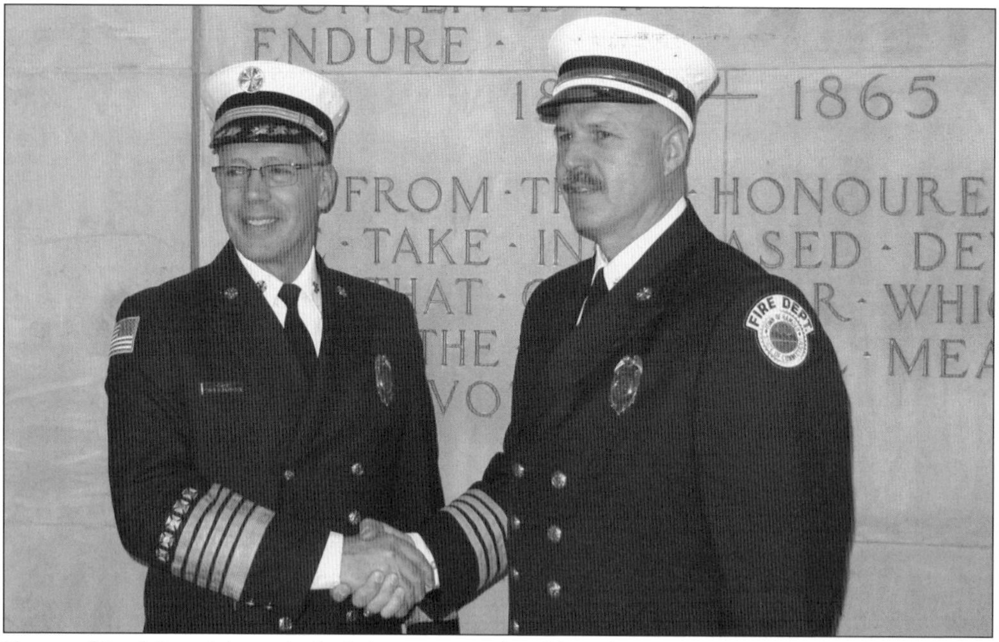

Deputy Chief Gary Merwede (right) is congratulated by Chief Berardesca following his January 2014 swearing-in ceremony in the town hall rotunda. He succeeded Deputy Chief Robert Surprise (retired). On the department since 1994, with the last 10 years as a company officer, Merwede served in every elected position in Local 2687. He was local president from 2005 until the end of 2011, whereupon he was unanimously voted president emeritus. (Robert Mordecai.)

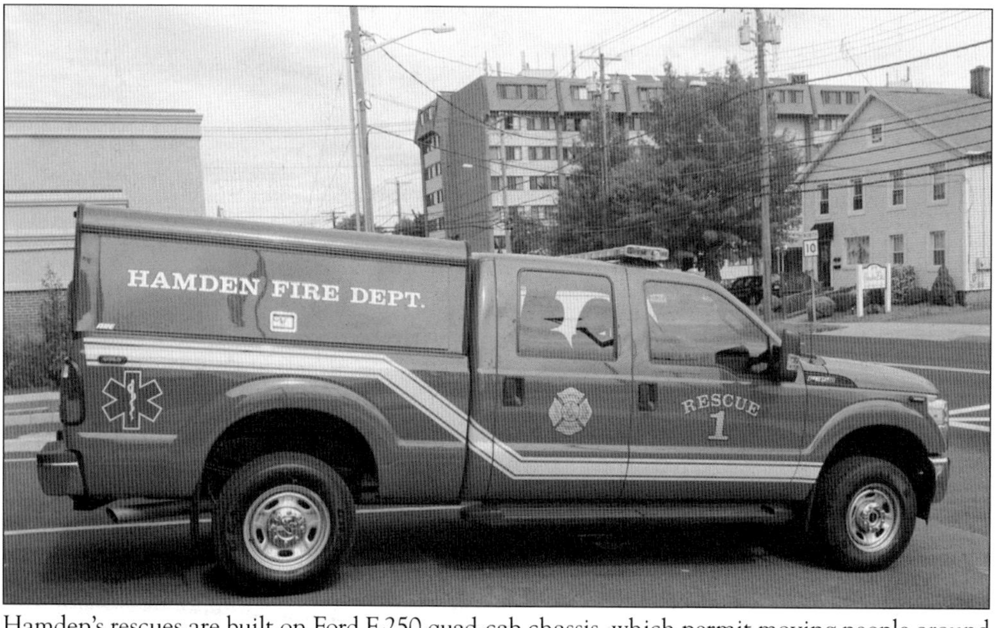

Hamden's rescues are built on Ford F-250 quad-cab chassis, which permit moving people around when necessary. The rear compartments have slide-out trays for fire and medical gear. Three identical vehicles are used for two rescue companies. With rescues receiving many thousands of fire and EMS calls annually, rotation of the three vehicles between both rescue companies allows for more even wear and tear, simplifying preventative maintenance. (HFRA.)

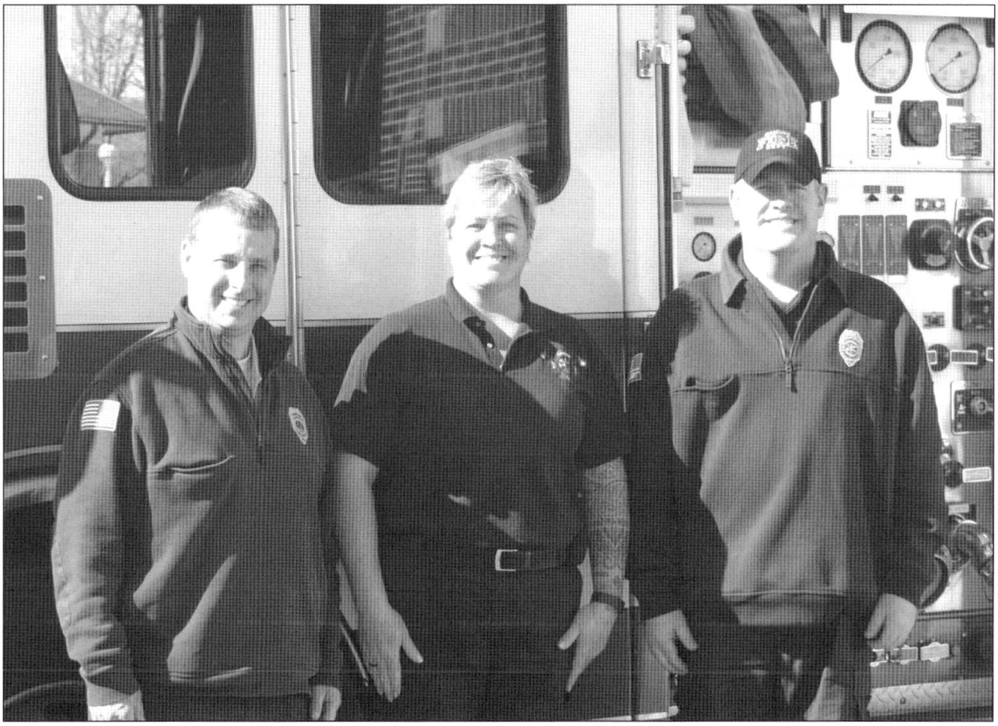

Firefighter Kim Talmadge (center) is flanked by Capt. Gregory Bannon (left) and Firefighter Jay McCarthy in this 2016 photograph. Firefighter Talmadge, Hamden's third female career firefighter, is assigned to Mt. Carmel Station 5, where Cheryl Flynn became Hamden's first female volunteer firefighter in 1974. Kerry Castracane was appointed Hamden's first female career firefighter in 1987. Firefighter/paramedic Roberta Angiletta followed in 1994. Both women served for more than 20 years. (David Johnson.)

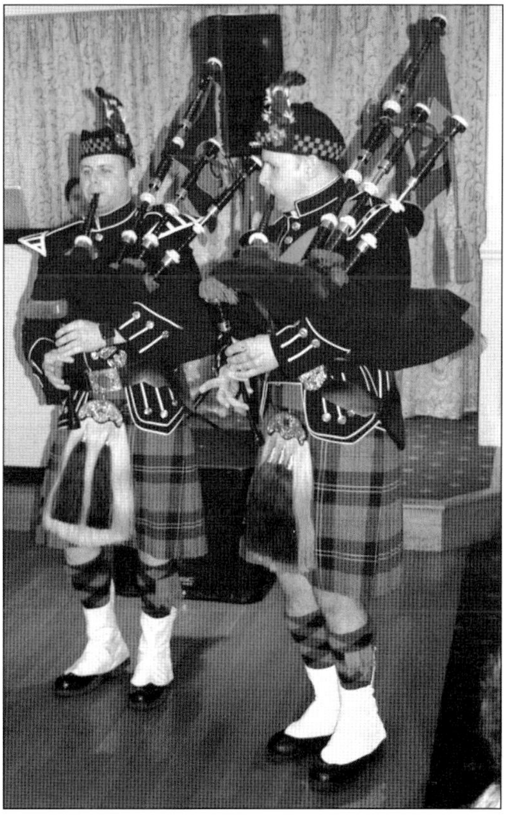

A major highlight of Local 2687's annual dinner is the opening ceremony performed by Capt. Edwin Evers (right) and Lt. Michael Dowling, both members of the Connecticut Firefighters Pipes and Drums. Founded in 1992 as a direct result of the 1990 accident that killed Firefighters Howard Hughes and Heriberto Rivera of Waterbury, the band performs 60 to 80 times annually at Connecticut fire department ceremonial functions and funerals. (HFRA.)

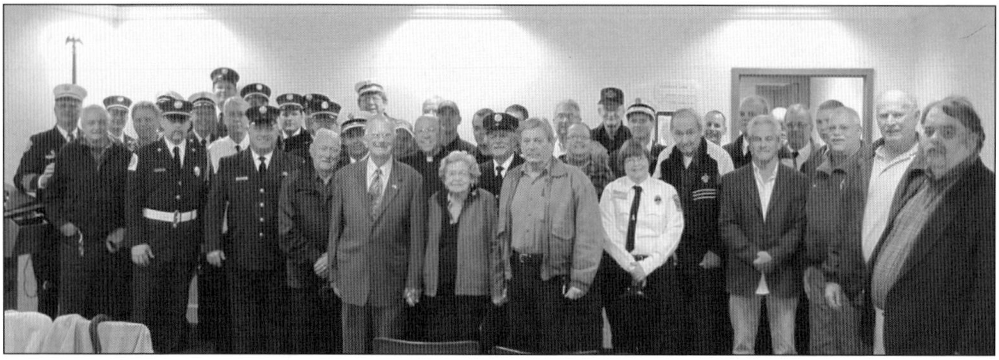

On December 10, 1927, Volunteer Firefighter Edward Meegan of Centerville was responding to an alarm on Engine 4 when the pumper flipped over in a freak accident. He died of his injuries the next day. Pictured here are the many active, retired, volunteer, and career Hamden firefighters who gathered at the Hamden Government Center to honor the memory of Firefighter Meegan on the 85th anniversary of his last alarm. Meegan's goddaughter Inez Meegan Shea, flanked by her husband Philip (left) and son Philip Jr., was presented with a mayoral proclamation making December 10, 2012, Edward Meegan Day in the town of Hamden. (Robert Mordecai.)

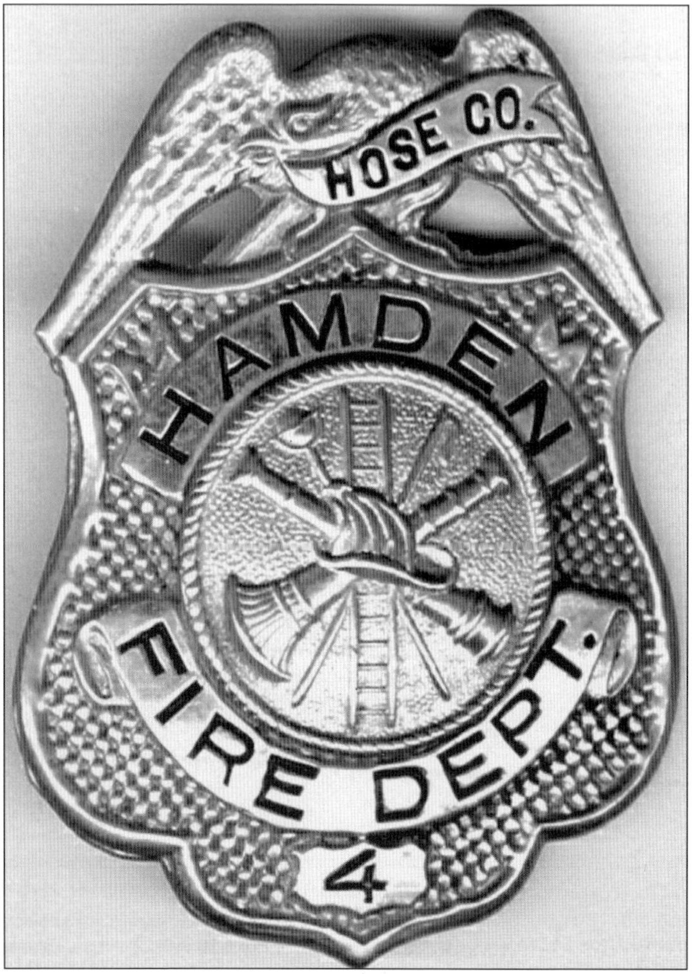

Inez Meegan Shea presented the Hamden Fire Retirees Association with her godfather's Centerville Co. 4 badge, which had been in her family since his tragic death. The badge will be kept in the HFRA's archives and future museum. (HFRA.)

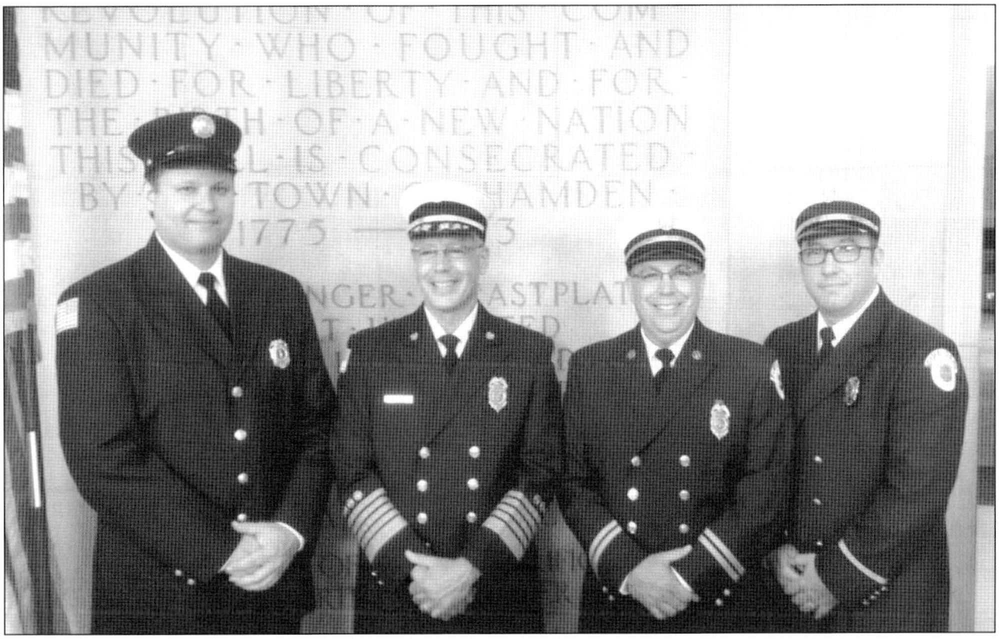

Swearing in new firefighters and newly promoted officers has become a department-wide celebration in recent years, with family members and scores of active and retired members attending ceremonies in the town hall rotunda. Following a May 2015 swearing-in ceremony, Chief David A. Berardesca (second from left) poses with, from left to right, newly promoted Lt. Daryl Osiecki, Battalion Chief Richard Otlowski, and Capt. Hugh O'Callaghan. (HFRA.)

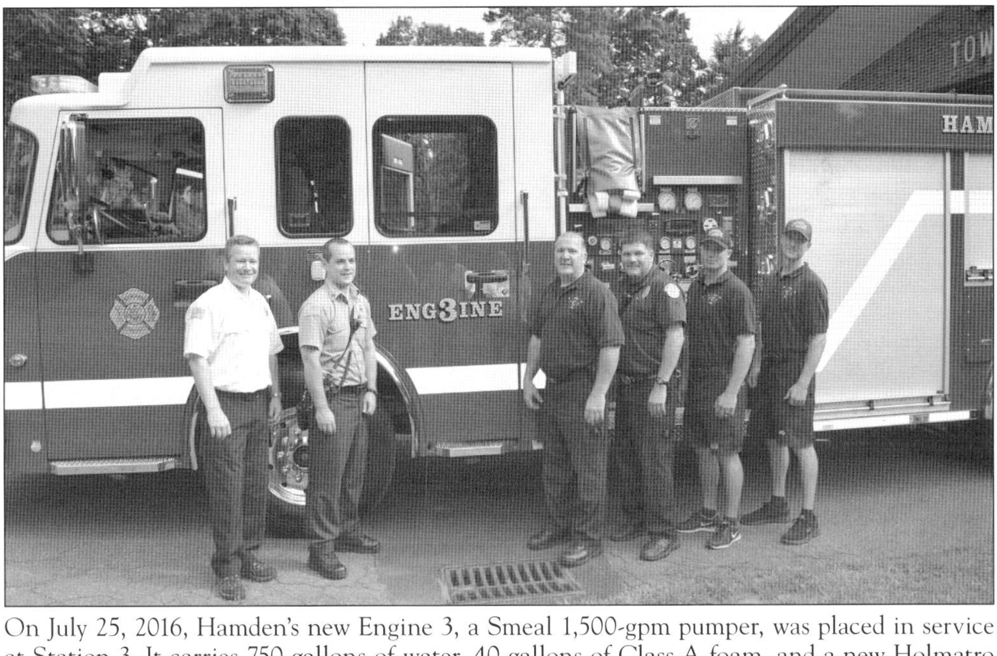

On July 25, 2016, Hamden's new Engine 3, a Smeal 1,500-gpm pumper, was placed in service at Station 3. It carries 750 gallons of water, 40 gallons of Class-A foam, and a new Holmatro hydraulic rescue tool. From left to right are station personnel Capt. Brion Tierney, Lt. Richard Lennon, and Firefighters Kurt Vogt, David Beaton, Steven Postemsky, and Brian Gagnon. Vogt is currently president of Local 2687. (HFRA.)